T0305838

Solutions Manual
for
Guide to Energy Management,
Eighth Edition

Solutions Manual
for
Guide to Energy Management, Eighth Edition

Klaus-Dieter E. Pawlik

Routledge
Taylor & Francis Group
LONDON AND NEW YORK

Published 2020 by River Publishers
River Publishers
Alsbjergvej 10, 9260 Gistrup, Denmark
www.riverpublishers.com

Distributed exclusively by Routledge
4 Park Square, Milton Park, Abingdon, Oxon OX14 4RN
605 Third Avenue, New York, NY 10017, USA

Solutions Manual for Guide to Energy Management, Eighth Edition
By Klaus-Dieter E. Pawlik

First published by Fairmont Press in 2016.

Routledge is an imprint of the Taylor & Francis Group, an informa business

ISBN 0-88173-771-2 (The Fairmont Press, Inc. Print Version)
ISBN 978-8-7702-2451-2 (print)
ISBN 978-8-7702-2331-7 (online)
ISBN 978-1-0031-7751-7 (ebook master)

While every effort is made to provide dependable information, the publisher, authors, and editors cannot be held responsible for any errors or omissions.

The views expressed herein do not necessarily reflect those of the publisher.

Table of Contents

Chapter 1

Introduction to Energy Management

Problem: For your university or organization, list some energy management projects that might be good "first ones," or early selections.

Solution: Early projects should have a rapid payback, a high probability of success, and few negative consequences (increasing/decreasing the air-conditioning/heat, or reducing lighting levels).

Examples:
Switching to a more efficient light source (especially in conditioned areas where one not only saves with the reduced power consumption of the lamps but also from reduced refrigeration or air-conditioning load).

Repairing steam leaks. Small steam leaks become large leaks over time.

Insulating hot fluid pipes and tanks.

Install high efficiency motors.

And many more

Problem: Again for your university or organization, assume you are starting a program and are defining goals. What are some potential first-year goals?

Solution: Goals should be tough but achievable, measurable, and specific.

Examples:
Total energy per unit of production will drop by 10 percent for the first and an additional 5 percent the second.

Within 2 years all energy consumers of 5 million British thermal units per hour (Btuh) or larger will be separately metered for monitoring purposes.

Each plant in the division will have an active energy management program by the end of the first year.

All plants will have contingency plans for gas curtailments of varying duration by the end of the first year.

All boilers of 50,000 lbm/hour or larger will be examined for waste heat recovery potential the first year.

Problem: Perform the following energy conversions and calculations:

a) A spherical balloon with a diameter of ten feet is filled with natural gas. How much energy is contained in that quantity of natural gas?

b) How many Btu are in 200 therms of natural gas? How many Btu in 500 gallons of 92 fuel oil?

c) An oil tanker is carrying 20,000 barrels of #2 fuel oil. If each gallon of fuel oil will generate 550 kWh of electric energy in a power plant, how many kWh can be generated from the oil in the tanker?

d) How much coal is required at a power plant with a heat rate of 10,000 Btu/kWh to run a 6 kW electric resistance heater constantly for 1 week (16 8 hours)?

e) A large city has a population which is served by a single electric utility which burns coal to generate electrical energy. If there are 500,000 utility customers using an average of 12,000 kWh per year, how many tons of coal must be burned in the power plants if the heat rate is 10,500 Btu/kWh?

f) Consider an electric heater with a 4,500 watt heating element. Assuming that the water heater is 98% efficient, how long will it take to heat 50 gallons of water from 70 degree F to 140 degree F?

Solution:

   a) V = 4/3 (PI) P
         = 4/3 × 3.14 × 5^3
            523.33 ft^3

      E = V × 1,000 Btu/cubic foot of natural gas
         = 523.33 ft^3 X 1,000 Btu/ft^3
         = *523,333 Btu*

   b) E = 200 therms × 100, 000 Btu/therm of natural gas
         = *20,000,000 Btu*
      E = 500 gallons × 140,000 Btu/gallon of #2 fuel oil
            70,000,000 Btu

   c) E = 20,000 barrels × 42 gal./barrel × 550 kWh/gal.
            4.6E+08 kWh

   d) V = 10,000 Btu/kWh × 6 kW × 168 h/25,000,000
            Btu/ton coal
         = *0.40 tons of coal*

   e) V = 500,000 cus. × 12,000 kWh/cus. × 10,500
            Btu/kWh × I ton/25,000,000 Btu
         = *2,520,000 tons of coal*

   f) E = 50 gal. × 8.34 lbm/gal. × (140F - 70F) ×
            1 Btu/F/lbm
         = 29,190 Btu
         = 29,190 Btu/3,412 Btu/kWh
         = 8.56 kWh
         = 8.56 kWh/4.5 kW/0.98
         = *1.94 h*

Problem: If you were a member of the upper level management in charge of implementing an energy management program at your university or organization, what actions would you take to reward participating individuals and to reinforce commitment to energy management?

Solution: The following actions should be taken to reward individuals and reinforce commitment to energy management:

Develop goals and a way of tracking their progress.

Develop an energy accounting system with a performance measure such as Btu/sq. ft or Btu/unit.

Assign energy costs to a cost center, profit center, an investment center or some other department that has an individual responsibility for cost or profit.

Reward (with a monetary bonus) all employees who control cost or profit relative to the level of cost or profit. At the risk of being repetitive, note that the level of cost or profit should include energy costs.

Problem: A person takes a shower for ten minutes. The water flow rate is three gallons per minute, the temperature of the shower water is 110 degrees E Assuming that cold water is at 65 degrees F, and that hot water from a 70% efficient gas water heater is at 140 degrees F, how many cubic feet of natural gas does it take to provide the hot water for the shower?

Solution: E = 10 min × 3 gal./min × 8.34 lbm/gal ×
                       (110 F - 65 F) × 1 Btu/lbm/F
                  = 11,259 Btu

              V = 11,259 Btu × 1 cubic foot/1,000 Btu/0.70
                  = *16.08 cubic feet of natural gas*

Problem: An office building uses 1 Million kWh of electric energy and 3,000 gallons of #2 fuel oil per year. The building has 45,000 square feet of conditioned space. Determine the Energy Use Index (EUI) and compare it to the average EUI of an office building.

Solution:

$$E(\text{elect.}) = 1{,}000{,}000 \text{ kWh/yr.} \times 3{,}412 \text{ Btu/kWh}$$
$$= 3{,}412{,}000{,}000 \text{ Btu/yr.}$$

$$E(\#2 \text{ fuel}) = 3{,}000 \text{ gal./yr.} \times 140{,}000 \text{ Btu/gal.}$$
$$= 420{,}000{,}000 \text{ Btu/yr.}$$

$$E = 3{,}832{,}000{,}000 \text{ Btu/yr.}$$

$$\text{EUI} = 3{,}832{,}000{,}000 \text{ Btu/yr.}/45{,}000 \text{ sq. ft}$$

$$= \textit{85,156 Btu/sq. ft/yr. which is}$$
$$\textit{less than the average office building}$$

Problem: The office building in Problem 1.6 pays $65,000 a year for electric energy and $3,300 a year for fuel oil. Determine the Energy Cost Index (ECI) for the building and compare it to the ECI for an average building.

Solution: ECI = ($65,000 + $3,300)/45,000 sq. ft
          = *$1.52/sq. ft/yr.*
          which is greater than the average building

Problem: As a new energy manager, you have been asked to pre-
dict the energy consumption for electricity for next month
(February). Assuming consumption is dependent on units
produced, that 1,000 units will be produced in February,
and that the following data are representative, determine
your estimate for February.

Month	Units produced	Consumption (kWh)	Average (kWh/unit)	
January	600	600	1.00	
February	1,500	1,200	0.80	
March	1,000	800	0.80	
April	800	1,000	1.25	
May	2,000	1,100	0.55	
June	100	700	7.00	Vacation month
July	1,300	1,000	0.77	
August	1,700	1,100	0.65	
September	300	800	2.67	
October	1,400	900	0.64	
November	1,100	900	0.82	
December	200	650	3.25	1-week shutdown
January	1,900	1,200	0.63	

Given: (see table above)

Solution: First, since June and December have special circumstances,
we ignore these months. We then run a regression to find
the slope and intercept of the process model. We assume
that with the exception of the vacation and the shutdown
that nothing other then the number of units produced
affects the energy used. Another method of solving this
problem may assume that the weather and temperature
changes also affects the energy use.

Month	Units produced	Consumption (kWh)	Average (kWh/unit)
January	600	600	1.00
February	1,500	1,200	0.80
March	1,000	800	0.80
April	800	1,000	1.25
May	2,000	1,100	0.55
July	1,300	1,000	0.77
August	1,700	1,100	0.65
September	300	800	2.67
October	1,400	900	0.64
November	1,100	900	0.82
January	1,900	1,200	0.63

From the ANOVA table, we see that if this process is modeled linearly the equation describing this is as follows:

kWh (1,000 units) = 623 + 0.28 × kWh/unit produced

= *899 kWh*

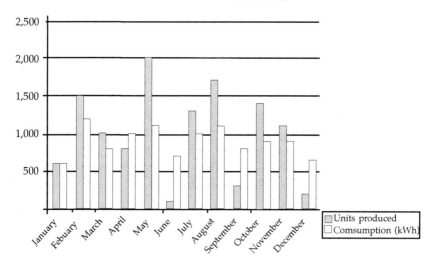

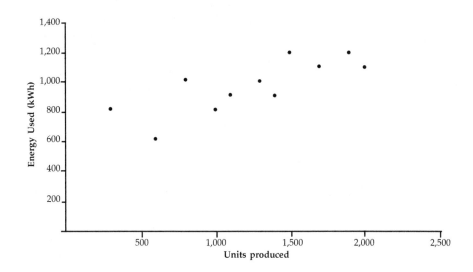

SUMMARY OUTPUT

Regression Statistics

Multiple R	0.795822426
R Square	0.633333333
Adjusted R Square	0.592592593
Standard Effort	118.6342028
Observations	11

ANOVA

	df	SS	MS	F	Significance F
Regression	1	218787.9788	218787.9	15.54545	0.00339167
Residual	9	126666.6667	14074.07		
Total	10	345454.5455			

	Coefficients	Standard Error	t Stat	P-value	Lower 95%	Upper 95%	Lower 95. 0%	Upper 95.0%
Intercept	623.1884058	93.46296795	6.667759	9.19E-05	411.7603222	834.616489	411.760322	834.6164893
X Variable 1	0,275362319	0.06993977	3.942772	0.003392	0,117373664	0.43335097	0.11737366	0.433350974

Problem: For the same data as given in Problem 1.8, what is the fixed energy consumption (at zero production, how much energy is consumed and for what is that energy used)?

Solution: *By* looking at the regression run for problem 1.8 (see ANOVA table), we can see the intercept for the process in question. This intercept is probably the best estimate of the fixed energy consumption:

       623 kWh.

       This energy is probably used for space conditioning and security lights.

Problem: Determine the cost of fuel switching, assuming there were 2,000 cooling degree days (CDD) and 1,000 units produced in each year.

Given: At the Gator Products Company, fuel switching caused an
increase in electric consumption as follows:

	Expected energy consumption	Actual energy consumption after switching fuel
Electric/CDD	75 million Btu	80 million Btu
Electric/units of production	100 million Btu	115 million Btu

The base year cost of electricity is $15 per million Btu, while this year's cost is $18 per million Btu.

Solution: Cost variance   = $18/million Btu - $15/million Btu
                              = $3/million Btu

Increase cost due to cost variance
    = Cost variance × Total Actual Energy Use
    = ($3/million Btu) × ((80 million Btu/CDD) ×
    (2,000 CDDs) + (115 million Btu/unit) × (1,000
units))
    = $825,000

CDD electric variance
    = 2,000 CDD × (80 - 75) million Btu/CDD
    = 10,000 million Btu

Units electric variance
    = 1,000 units × (115 - 100) million Btu/unit
    = 15,000 million Btu

Increase in energy use
  = CDD electric variance + Units electric variance
  = 10,000 million Btu + 15,000 million Btu
  = 25,000 million Btu

Increase cost due to increased energy use
  = (Increase in energy use) × (Base cost of electricity)
  = 25,000 billion Btu × $15/million Btu
  = $375,000

Total cost of fuel switching
  = Increase cost due to increased energy use
    + Increased cost due to cost variance
  = $375,000 + $825,000
  = *$1,200,000*

Chapter 2

The Energy Audit Process: An Overview

Problem: Compute the number of heating degree days (HDD) associated with the following weather data.

		Tempera-ture (degrees F)	Number of hours	65F -Temperature (degrees F)	Hours × dT
Given:	**Time Period**				
	Midnight - 4:00 AM	20	4	45	180
	4:00 AM - 7:00 AM	15	3	50	150
	7:00 AM - 10:00 AM	18	3	47	141
	10:00 AM - Noon	22	2	43	86
	Noon - 5:00 PM	30	5	35	175
	5:00 PM - 8:00 PM	25	3	40	120
	8:00 PM - Midnight	21	4	44	176
					1,028

Solution: From the added columns in the given table, we see that the number of hours times the temperature difference from 65 degrees F is 1,028 F-hours. Therefore, the number of HDD can be calculated as follows:

$$\text{HDD} = 1,028 \text{ F-hours}/24 \text{ h/day}$$
$$= \textbf{42.83 degree-days}$$

15

Problem: Select a specific type of manufacturing plant and describe the kinds of equipment that would likely be found in such a plant.

List the audit data that would need to be collected for each piece of equipment.

What particular safety aspects should be considered when touring the plant?

Would any special safety equipment or protection be required?

Solution: The following equipment could be found in a wide variety of manufacturing facilities:

Equipment	Audit data
Heaters	Power rating
	Use characteristics (annual use, used in conjunction with what other equipment, how is the equipment used?)
Boilers	*Power rating*
	Use characteristics
	Fuel used
	Air-to-fuel ratio
	Percent excess air
Air-conditioners	Power rating
Chillers	Efficiency
Refrigeration	Cooling capacity
	Use characteristics
Motors	Power rating
	Efficiency
	Use characteristics
Lighting	Power rating
	Use characteristics
Air-compressors	Power rating
	Use characteristics
	Efficiency
	Various air pressures
	An assessment of leaks

Specific process equipment for example for a metal furniture plant one may find some sort of electric arc welders for which one would collect its power rating and use characteristics.

The following include a basic list of some of the safety precautions that may be required and any safety equipment needed:

Safety precaution   *Safety equipment*

As a general rule of thumb the auditor should never touch anything: just collect data. If a measurement needs to be taken or equipment manipulated ask the operator.

Beware of rotating machinery
Beware of hot machinery/pipes   *Asbestos gloves*
Beware of live circuits   *Electrical gloves*

Have a trained electrician take any electrical measurements

Avoid working on live circuits, if possible
Securely lock and tag circuits and switches in the off/open position before working on a piece of equipment
Always keep one hand in your pocket while making measurements on live circuits to help prevent accidental electrical shocks.
When necessary, wear a full face respirator mask with adequate filtration particle size.
Use activated carbon cartridges in the mask when working around low concentrations of noxious gases. Change cartridges on a regular basis.
Use a self-contained breathing apparatus for work in toxic environments.
Use foam insert plugs while working around loud machinery to reduce sound levels by nearly 30 decibels (in louder environments hearing protection rated at higher noise levels may be required)

Always ask the facility contact about special safety precautions or equipment needed. Additional information can be found in OSHA literature.

For our metal furniture plant:
Avoid looking directly   *Tinted safety goggles*
at the arc of the welders

Problem: Section 2.1.2 of the *Guide to Energy Management* provided a
              list of energy audit equipment that should be used. How-
              ever, this list only specified the major items that might be
              needed. In addition, there are a number of smaller items
              such as hand tools that should also be carried. Make a list
              of these other items, and give an example of the need for
              each item.

              How can these smaller items be conveniently carried to the
              audit?

              Will any of these items require periodic maintenance or
              repair?

              If so, how would you recommend that an audit team keep
              track of the need for this attention to the operating condi-
              tion of the audit equipment?

Solution: Smaller useful audit equipment may include:
                  A flashlight
                  Extra batteries
                  A hand-held tachometer
                  A clamp-on ammeter
                  Recording devices

              These smaller items can be conveniently be carried in a
              tool box.

              As with most equipment, these items will require periodic
              maintenance. For example, the flashlight batteries and light
              bulbs will have to be changed.

              For these smaller items, one could probably just include
              the periodic maintenance as part of a pre-audit checklist.
              For items that require more than just cursory maintenance,
              one could include the item in their periodic maintenance
              system.

Problem: Section 2.2 of the *Guide to Energy Management* discussed the point of making an inspection visit to a facility at several different times to get information on when certain pieces of equipment need to be turned on and when they are unneeded. Using your school classroom or office building as a specific example, list some of the unnecessary uses of lights, air conditioners, and other pieces of equipment. How would you recommend that some of these uses that are not necessary be avoided? Should a person be given the responsibility of checking for this unneeded use? What kind of automated equipment could be used to eliminate or reduce this unneeded use?

Solution: Typically, one could visit a university at night and observe that the lights of classrooms are on even at midnight when no one is using the area. One idea would be to make the security force responsible for turning off non-security lights when they make their security tours at night. A better idea may be to install occupancy sensors so that the lights are on only when the area is in use. An additional benefit of an occupancy sensors could be security; many thieves or vandals would be startled when lights come on.

Problem: An outlying building has a 25 kW company-owned trans-
              former that is connected all the time. A call to a local
              electrical contractor indicates that the core losses from
              comparable transformers are approximately 3% of rated
              capacity. Assume that the electrical costs are ten cents
              per kWh and $10/kW/month of peak demand, that the
              average building use is ten hours/month, and that the
              average month has 720 hours. Estimate the annual cost
              savings from installing a switch that would energize the
              transformer only when the building was being used.

Given: Transformer power use   25   kW
              Core losses   3%
              Electrical energy cost $0.10   /kWh
              Demand charge $10/kW   /month
              Building utilization   10 hrs   /mo
              Hours in a month   720 hrs   /mo
              Months in a year   12 mo   /yr

Solution: The energy savings (ES) from installing a switch that
              would energize the transformer only when the building
              was being used can be calculated as follows:

              ES =   (Percentage of core losses) (Transformer power
                      use)(Hours in a month - Building utilization)
                      (Months in a year)
                  =   3% × 25 kW × 720 - 10) hrs/mo × 12 mo/yr
                  =   6,390 kWh/yr

              Since we do not expect the monthly peak demand to be
              reduced by installing this switch, the only savings will
              come from energy savings. Therefore, annual savings (AS)
              can be calculated as follows:

              AS =   ES × Electrical energy cost
                  =   6,390 kWh/yr × $ 0.10/kWh
                  =   *$ 639/yr*

Chapter 3

Understanding Energy Bills

Problem: By periodically turning off a fan, what is the total dollar savings per year to the company?

Given: In working with Ajax Manufacturing Company, you find six large exhaust fans are running constantly to exhaust general plant air (not localized heavy pollution). They are each powered by 30-hp electric motors with loads of 27 kW each. You find they can be turned off periodically with no adverse effects. You place them on a central timer so that each one is turned off for 10 minutes each hour. At any time, one of the fans is off, and the other five are running. The fans operate 10 h/day, 250 days/year. Assume the company is on the rate schedule given in Figure 3-10. Neglect any ratchet clauses. The company is on service level 3 (distribution service). (There may be significant HVAC savings since conditioned air is being exhausted but ignore that for now.)

Solution: Demand charge

On-peak $12.22/kW/mo	June-October	5 months/year
Off-peak $4.45/kW/mo	November-May	7 months/year

Energy charge
For first two million kWh $0.03431/kWh
All kWh over two million $0.03010/kWh

Assumptions (and possible explanations)
Assume the company uses well over two million kWh per month
The fuel cost adjustment is zero, since the utility's fuel cost is at the base rate.
There is no sales tax since the energy can be assumed to be used for production

The power factor is greater than 0.8

No franchise fees since the company is outside any municipality

The demand savings (DS) can be calculated as follows:

$$DS = [(DC \text{ on peak}) \times (N \text{ on peak}) + (DC \text{ off peak}) \times (N \text{ off peak})] \times DR$$

where,

DC = Demand charge for specified period
N = Number of months in a specified period
DR = Demand reduction, 27 kW since a motor using this amount is always turned off with the new policy

Therefore,

$$DS = [(\$12.22/kW/mo) \times (5 \text{ mo/yr}) + (\$4.45/kW/mo) \times (7 \text{ mo/yr})] \times 27 \text{ kW} = \$2,491/yr$$

The energy savings (ES) can be calculated as follows:

$$ES = (EC >2 \text{ million}) \times (10 \text{ h/day}) \times (250 \text{ day/yr}) \times DR$$

where

EC = Marginal energy charge

Therefore,

$$ES = (\$0.03010/kWh) \times (10 \text{ h/day}) \times (250 \text{ day/yr}) \times 27 \text{ kW}$$
$$= \$2,032/yr$$

Finally, the total annual savings (TS) can be calculated as follows:

$$TS = DS + ES$$
$$= \$4,523/yr$$

Additional Considerations

How much would these timers cost?

How much would it cost to install these timers? Or an alternate control system?

Does cycling these fans on and off cause the life of the fan motors to decrease?

What would the simple payback period be?

Net present value?

Internal rate of return?

Problem: What is the dollar savings for reducing demand by 100 kW in the off-peak season?

If the demand reduction of 100 kW occurred in the peak season, what would be the dollar savings (that is, the demand in June through October would be reduced by 100 kW)?

Given: A large manufacturing company in southern Arizona is on the rate schedule shown in Figure 3-10 (service level 5, secondary service). Their peak demand history for last year is shown below. Assume they are on the 65% ratchet clause specified in Figure 3-10. Assume the high month was July of the previous year at 1,150 kW.

Month	Demand (kW)	Month	Demand
Jan	495	Jul	1100
Feb	550	Aug	1000
Mar	580	*Sep*	*900*
Apr	600	Oct	600
May	610	Nov	500
Jun	*900*	Dec	515

note italics indicates on-peak season

Solution: Demand charge
On-peak $13.27/kW/mo   June-October   5 months/year
Off-peak $4.82/kW/mo   November-May   7 months/year

Ratchet clause
Dpeak = max (actual demand corrected for pf, 65% of the highest on-peak season demand corrected for pf)

Assumptions (and possible explanations)
Assume the company uses well over two million kWh per month.

The fuel cost adjustment is zero, since the utility's fuel cost is at the base rate.

There is no sales tax since the energy can be assumed to be used for production.

The power factor is greater than 0.8.

No franchise fees since the company is outside any municipality.

Estimated next year with a 100 kW decrease in the off-peak season

Month	Demand (kW)	Ratchet	Dollar savings
Jan	395	747.5	0
Feb	450	747.5	0
Mar	480	747.5	0
Apr	500	747.5	0
May	510	747.5	0
Jun	900	747.5	0
Jul	1100	715	0
Aug	1000	715	0
Sep	900	715	0
Oct	600	715	0
Nov	400	715	0
Dec	415	715	0
			0

Therefore, you would not save any money by reducing the peak demand in the off-season. This non-savings is due to the ratchet and the degree of unevenness of demand.

Estimated next year with a 100 kW decrease in the on-peak season

Month	Demand (kW)	Ratchet	Dollar savings
Jan	495	747.5	0
Feb	550	747.5	0
Mar	580	747.5	0
Apr	600	747.5	0
May	610	747.5	0
Jun	*800*	*747.5*	*$1,327*
Jul	*1000*	*650*	*$1,327*
Aug	900	650	$1,327
Sep	*800*	*650*	*$1,327*
Oct	*500*	*650*	*$863*
Nov	500	650	$313
Dec	515	650	$313

The first year they would save: $6,797

Every year after that they would save the following:

$$\text{Savings} = 65\% \times 100 \text{ kW} \times 7 \text{ mo/yr} \times \$4.82/\text{kW/mo}$$
$$+ 100 \text{ kW} \times 5 \text{ mo/yr} \times \$13.27/\text{kW/mo}$$
$$= \$8,828/yr$$

Problem: Use the data found in Problem 3.2. How many months would be ratcheted, and how much would the ratchet cost the company above the normal billing?

Solution: Assuming that the 100 kW reduction is not made

Month	Demand (kW)	Ratchet	Ratchet Cost
Jan	495	747.5	$1,217.05
Feb	550	747.5	$951.95
Mar	580	747.5	$807.35
Apr	600	747.5	$710.95
May	610	747.5	$662.75
Jun	900	747.5	$—
Jul	1100	715	$—
Aug	1000	715	$—
Sep	900	715	$—
Oct	600	715	$1,526.05
Nov	500	715	$1,036.30
Dec	515	715	$964.00

8 months would be ratcheted at a cost of $7,876.40

Problem: Calculate the savings for correcting to 80% power factor? How much capacitance (in kVARs) would be necessary to obtain this correction?

Given: In working with a company, you find they have averaged 65% power factor over the past year. They are on the rate schedule shown in Figure 3-10 and have averaged 1,000 kW each month. Neglect any ratchet clause and assume their demand and power factor are constant each month. Assume they are on transmission service (level 1).

Solution: <u>Demand Charge</u>
              On-peak \$10.59/kW/mo   June-October   5 months/year
              Off-peak \$3.84/kW/mo   November -May   7 months/year

              Billed Demand   = Actual Demand × (base pf/actual pf)
                               = 1000 kW × 0.8/0.65
                               = 1231 kW

              pf correction savings   = 231 kW × (5 mo/yr × \$10.59/kW/mo
                                         + 7 mo/yr × \$3.82/kW/mo)
                                       = *\$18,422/yr*

                          pf   = cos(theta) = 0.65
                       theta   = 0.86321189 radians
                 kVAR initial   = 1000 kW × tan (0. 86)
                               = 1169 kVAR

                          pf   = cos(theta) = 0.8
                       theta   = 0.643501109 radians
                 kVAR initial   = 1000 kW × tan (0. 86)
                                 750 kVAR

       capacitor size needed   = *419 kVAR*

Also, using a pf correction table for 0.65 => 0.80:

                       kVAR   = (0.419) × (1000 kW)
                               = *419 kVAR*

Problem: How much could they save by owning their own trans-
             formers and switching to service level 1?

Given: A company has contacted you regarding their rate sched-
             ule. They are on the rate schedule shown in Figure 3-10,
             service level 5 (secondary service), but are near transmis-
             sion lines and so can accept service at a higher level (ser-
             vice level 1) if they buy their own transformers. Assume
             they consume 300,000 kWh/month and are billed for 1,000
             kW each month. Ignore any charges other than demand
             and energy.

Solution:

Service level 1 (proposed)
Demand Charge

On-peak	$10.59 /kW/mo	June-Oct.	5 months/year
Off-peak	$3.84 /kW/mo	Nov.-May	7 months/year

Energy Charge

For first two million kWh	$0.03257 /kWh
All kWh over two million	$0.02915 /kWh

Service level 5 (present)
Demand Charge

On-peak	$13.27 /kW/mo	June-Oct.	5 months/year
Off-peak	$4.82 /kW/mo	Nov.-May	7 months/year

Energy Charge

For first two million kWh	$0.03528 /kWh
All kWh over two million	$0.03113 /kWh

Rate savings:
Demand Charge

On-peak	$2.68 /kW/mo	June-Oct.	5 months/year
Off-peak	$0.98 /kW/mo	Nov.-May	7 months/year

Energy Charge

For first two million kWh	$0.00271 /kWh
All kWh over two million	$0.00198 /kWh

ES = 300,000 kWh/mo × 12 mo/yr × $0.00271/kWh
   = $9,756 /yr
DS = 1,000 kW ($2.68/kW/mo×5 mo/yr+$0.98/kW/mo×7 mo/yr)
   = $20,260/yr
TS = *$30,016/yr*

Problem: What is the savings from switching from priority 3 to priority 4 rate schedule?

Given: In working with a brick manufacturer, you find for gas billing that they were placed on an industrial (priority 3) schedule (see Figure 3-12) some time ago. Business and inventories are such that they could switch to a priority 4 schedule without many problems. They consume 7,000 Mcf of gas per month for process needs and essentially none for heating.

Solution:

Priority 3 (present)

	Schedule	Rate	Monthly Cost (for 7,000 Mcf/mo)
First	1 ccf	$19.04	$19.04
Next	2.9 Mcf/mo	$5.490 /Mcf	$15.92
Next	7 Mcf/mo	$5.386 /Mcf	$37.70
Next	90 Mcf/mo	$4.372 /Mcf	$393.48
Next	100 Mcf/mo	$4.127 /Mcf	$412.70
Next	7800 Mcf/mo	$3.445 /Mcf	$23,426.00
Over	8000 Mcf/mo	$3.399 /Mcf	$————
Total present monthly cost:			$24,304.84

Total present annual cost: $291,658.12

Priority 4 (proposed)

	Schedule	Rate	Monthly Cost (for 7,000 Mcf/mo)
First 4,000 Mcf/mo or fraction thereof		$12,814	$12,814
	Next 4000 Mcf/mo	$3.16 /Mcf	$9,504
	Over 8000 Mcf/mo	$3.122 /Mcf	$————
Total present monthly cost:			$22,318.00

Total present annual cost: $267,816.00
Annual savings from switching: $23,842.12

Additional Considerations
What-if there exists a 20% probability that switching to the proposed rate schedule will disrupt production one more time a year for an hour?

Problem: Calculate the January electric bill for this customer.

Given: *A* customer has a January consumption of 140,000 kWh, a peak 15-minute demand during January of 500 kW, and a power factor of 80%, under the electrical schedule of the example in Section 3.6.
Assume that the fuel adjustment is:
$0. 01/kWh

Solution: Quantity Cost

Customer charge	$21.00	/mo	1 mo	$21
Energy charge	$0.04	/kWh	140,000 kWh	$5,600
Demand charge	$6.50	/kW/mo	500 kW	$3,250
Taxes	8%			
Fuel Adjustment	$0.01	/kWh	140,000 kWh	$1,400
			sub-total	$10,271
			tax	822
			total	*$11,093*

Problem: Compare the following residential time-of-use electric rate with the rate shown in Figure 3-6.

Given: Customer charge $8.22   /mo
           Energy charge $0.1230 /kWh   on-peak
                               $0.0489 /kWh   off-peak

This rate charges less for electricity used during off-peak hours—about 80% of the hours in a year—than it does for electricity used during on-peak hours.

Solution: Each of the rates have a different on-peak period. However, if we assume that no matter which rate schedule is used that 80% of the energy is used off-peak, then average cost per kWh can be calculated as follows:

AC = (Off-peak percentage of energy use)(Off-peak energy cost) +
       (1 - off-peak percentage of energy use)(On-peak energy cost)

Therefore, the average cost per kWh with the above schedule is:

AC = (80%)($0.0489/kWh) + (1-80%)($0,123/kWh)
    = *$0.06372/kWh*

And the average cost per kWh with the schedule in figure 3-6 is:

AC = (80%)($0.0058/kWh) + (1-80%)($0,10857/kWh)
    = *$0.02635/kWh*

Problem: What is the power factor of the combined load?

If they added a second motor that was identical to the one they are presently using, what would their power factor be?

Given: A small facility has 20 kW of incandescent lights and a 25
kW motor that has a power factor of 80%.

Solution: The lamp:

           20 kW

The motor:

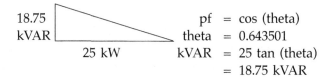

$$pf = \cos(\theta)$$
$$\theta = 0.643501$$
$$kVAR = 25\tan(\theta)$$
$$= 18.75 \text{ kVAR}$$

Combined:

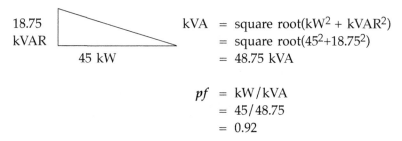

$$kVA = \text{square root}(kW^2 + kVAR^2)$$
$$= \text{square root}(45^2 + 18.75^2)$$
$$= 48.75 \text{ kVA}$$

$$pf = kW/kVA$$
$$= 45/48.75$$
$$= 0.92$$

Combined:

$$kVA = \text{square root}(kW^2 + kVAR^2)$$
$$= \text{square root}(70^2 + 37.5^2)$$
$$= 79.41 \text{ kVA}$$

$$pf = kW/kVA$$
$$= 70/79.41$$
$$= \mathbf{0.88}$$

Problem: For the load curve shown below for Jones Industries, what is their billing demand and how many kWh did they use in that period?

Given: *A* utility charges for demand based on a 30-minute synchronous averaging period.

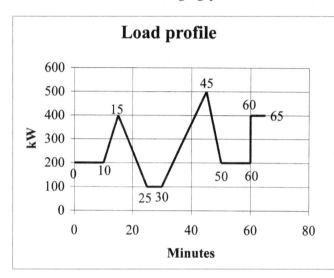

Minutes	kW
0	200
10	200
15	400
25	100
30	100
45	500
50	200
60	200
60	400
65	400

Solution: For the first 30 minutes: For the second 30 minutes:

Time (minutes)	average kW	Time (minutes)	average kW
10	200	15	300
5	300	5	350
10	250	10	200
5	100		

Weighted average: 216.67 kW   Weighted average: *275.00 kW*

Therefore, **275 kW** is the billed demand.

kWh = (216.67 kW)(0.5 hours) + (275 kW)(0.5 hours) +
        (400 kW)(5 minutes x 1 hour/60 minutes))
      = *279.17 kWh*

Problem: Based on the hypothetical steam rate in Figure 3-13, determine their steam consumption cost for the month?

Given: The Al Best Company has a steam demand of 6,500 lb/hr and a consumption of 350,000 lbs during the month of January.

Solution: <u>Steam consumption charge</u>

$3.50 /1000 lb. for the first 100, 000 lb of steam per month
$3.00 /1000 lb. for the next 400,000 lb of steam per month
$2.75 /1000 lb. for the next 500,000 lb of steam per month
$2.00 /1000 lb. for the next 1,000,000 lb of steam per month

$$
\begin{aligned}
\textit{Consumption cost} \ &= \ \$3.50/1{,}000 \ \text{lb} \times 100{,}000 \ \text{lb} \\
&\quad + \ \$3.00/1{,}000 \ \text{lb} \times 250{,}000 \ \text{lb} \\
&= \ \$350 + \$750 \\
&= \ \boldsymbol{\$1{,}100}
\end{aligned}
$$

Problem: What is Al's cost for chilled water in July?
             What was their Btuh (Btu/hour) equivalent for the average
             chilled water demand?

Given: Al Best also purchases chilled water with the rate schedule
             of figure 3-13. During the month of July, their chilled water
             demand was
                     485 tons and their consumption was
                     250,000 ton-hours

Solution: Chilled water demand charge:
             $2,500 /mo for the first   100 tons or any portion thereof
               $15 /mo/ton for the next 400 tons
               $12 /mo/ton for the next 500 tons
               $10 /mo/ton for the next 500 tons
                $9 /mo/ton for over   1500 tons

             <u>Chilled water consumption charge</u>:
             $0.069 /tonh   for the first 10,000 tonh/mo
             $0.060 /tonh   for the next 40,000 tonh/mo
             $0.055 /tonh   for the next 50,000 tonh/mo
             $0.053 /tonh   for the next 100,000 tonh/mo
             $0.051 /tonh   for the next 100,000 tonh/mo
             $0,049 /tonh   for the next 200,000 tonh/mo
             $0.046 /tonh   for the next 500,000 tonh/mo

             Demand cost = $2,500 + ($15/ton)(385 tons)
                          = $8,275

Consumption cost = $0.069   ×   10,000 tonh/mo +
                      $0.060   ×   40,000 tonh/mo +
                      $0.055   ×   50,000 tonh/mo +
                      $0.053   ×   100,000 tonh/mo +
                      $0.051   ×   50,000 tonh/mo
                   = $13,690

        Total bill   = $21,965

Average demand   = consumption × 12,000 Btuh/ton 744 hr/July
                  = 250,000 tonh × 12,000 Btuh/ton 744 hr/July
                  = *4,032,258 Btuh in July*

Chapter 4

Economic Analysis and Life Cycle Costing

Problem: *How* much can they spend on the purchase price for this project and still have a Simple Payback Period (SPP) of two years?

Using this figure as a cost, what is the return on investment (ROI), and the Benefit-Cost Ratio (BCR)?

Given: The Orange and Blue Plastics Company is considering an energy management investment which will save 2,500 kWh of electric energy at $0.08/kWh. Maintenance will cost $50 per year, and the company's discount rate is 12%.

Solution: Annual savings = annual kWh saved × electric energy
cost-maintenance cost
= 2,500 kWh/yr × $0.08/kWh - $50/yr
= $150/yr

Implementation cost = SPP × Annual savings
= 2 yrs × $150/yr
= *$300*

Since no life is given, assume the project continues forever. Therefore, use the highest n in the TMV tables: n = 360

$$P = A[P | A, i, N]$$
$$300 = 150 [P | A, i, 360]$$
$$2 = [P | A, i, 360]$$

From the TMV tables, we see that i ~ 50%. Therefore,
ROI = 50%

$$\begin{aligned} \text{BCR} &= \text{PV(benefits)}/\text{PV (costs)} \\ \text{PV (benefits)} &= A[P \mid A, i, N] \\[1em] &= \$150 \times [P \mid A, 12\%, 360 \text{ yr}] \\ &= \$150 \times 8.3333 \\ &= \$1,250 \\ \textit{BCR} &= \$1,250/\$300 \\ &= \textit{4.17} \end{aligned}$$

If N = 5 years:

we read from the TMV tables that the factor 2 falls between 40% and 50% tables with the factors 2.0352 and 1.7366 respectively. Therefore, to find a more precise percentage we linearly interpolate:

(50% - 40%)/(1.7366 - 2.0352) (x - 40%)/(2 - 2.0352)
Solving for X:

$$X = \textit{41.2 \%} = ROI$$

$$\begin{aligned} \text{BCR} &= \text{PV (benefits)}/\text{PV (costs)} \\ \text{PV (benefits)} &= A[P \mid A, i, N] \\ &= \$150 [P \mid A, 12\%, 5 \text{ yr}] \\ &= \$150 \times 3.6048 \\ &= \$541 \\ \textit{BCR} &= \$541/\$300 \\ &= \textit{1.80} \end{aligned}$$

Problem: Which model should she buy to have the lowest total monthly payment including the loan and the utility bill?

Given: A new employee has just started to work for Orange and Blue Plastics, and she is debating whether to purchase a manufactured home or rent an apartment. After looking at apartments and manufactured homes, she decides to buy one of the manufactured homes. The Standard Model is the basic model that costs $20,000 and has insulation and appliances that have an expected utility cost of $150 per month. The Deluxe Model is the energy efficient model that has more insulation and better appliances, and it costs $22,000. However, the Deluxe Model has expected utility costs of only $120/month. She can get a 10 year loan for 10% for the entire amount of either home.

Solution: Assume the 10% is the compounded annual percentage rate.

$$P = A\,[P\,|\,A,\ i,\ N]$$
$$= A\,[P\,|\,A,\ 10\%,\ 10\ \text{years}]$$
$$= A\,(6.1446)$$
$$A\ (\text{Standard}) = \$20{,}000/6.1446\ \text{yrs}$$
$$= \$3{,}254.89/\text{yr}$$
$$= \$271.24/\text{mo}$$
$$\textit{Monthly (Standard)} = \$271.24/\text{mo} + \$150/\text{mo}$$

$$= \mathbf{\$421.24/\textit{mo}}$$

$$A\ (\text{Standard}) = \$22{,}000/6.1446\ \text{yrs}$$
$$= \$3{,}580.38/\text{yr}$$
$$= \$298.36/\text{mo}$$
$$\text{Monthly (Deluxe)} = \$298.36/\text{mo} + \$120/\text{mo}$$

$$= \mathbf{\$418.36/\textit{mo}}$$

Therefore, if she buys the deluxe, she will have a slightly lower monthly cost

Problem: Determine the SPP, ROI, and BCR for this project:

Given: The Al Best Company uses a 10-hp motor for 16 hours per day, 5 days per week, 50 weeks per year in its flexible work cell. This motor is 85% efficient, and it is near the end of its useful life. The company is considering buying a new high efficiency motor (91% efficient) to replace the old one instead of buying a standard efficiency motor (86.4% efficient). The high efficiency motor cost $70 more than the standard model, and should have a 15-year life. The company pays $7 per kW per month and $0.06 per kWh. The company has set a discount rate of 10% for their use in comparing projects.

Solution: Assume the load factor (If) is 60%.

DR = If × Pm × 0.746 kW/hp × ((1/effs) - (1/effh))

where,

DR = Demand reduction
Pm = Power rating of the motor, 10-hp
effs = Efficiency of the standard efficiency motor, 86.4%
effh = Efficiency of the high efficiency motor, 91%

Therefore,

DR = 0.6 × 10 hp × 0.746 kW/hp × ((1/0.864) - (1/0.91))
   = 0.26 kW

DCR = DR × DC × 12 mo/yr

where,

DCR = Demand cost reduction
DC = Demand cost, $7/kW/mo

Therefore,

DCR = 0.26 kW × $7/kW/mo × 12 mo/yr
    = $22.00/yr

ES DR × 16 hr/day × 5 days/wk × 5 0 wk/yr

Therefore,

ES = 0.26 kW × 16 hr/day × 5 days/wk × 50 wk/yr
   = 1,047.5 KWh/yr

ECS = ES × EC

where,
  ECS = Energy cost savings
  EC = Energy cost, $0.06/kWh

Therefore,
  ECS = 1,047.5 kWh/yr × $0.06/kWh
      = $62.85/yr

Therefore, the annual cost savings (ACS) can be calculated as follows:

  ACS = DCS + ECS
      = $22.00/yr + $62.85/yr
      = $84.85/yr

  SPP = Cost premium ACS
      = $70.00/$84.85/yr
      = *0.825 yrs*

Additionally, the ROI can be found with looking up the following factor in the interest rate tables:

     P = A [P I A,i,N]
  $70 = $84.85/yr [P I A, ROI, 14 years]
 0. 825 = [P I A, ROI, 14 years]
  ROI = 121.2%

       BCR = PV(benefits)/PV (costs)
 PV (benefits) = A[P I A, i, N]
             = $85.85/yr [P I A, 10%, 14 yr]
             = $84.85 × 7.3667
             = $625
       BCR = $625/$70
             = *8.93*

Problem: Using the BCR measure, which project should the company select? Is the answer the same if Life Cycle Costs (LCC) are used to compare the projects?

Given: Craft Precision, Incorporated must repair their main air conditioning system, and they are considering two alternatives. (1) purchase a new compressor for $20,000 that will have a future salvage value of $2,000 at the end of its 15 year life; or (2) purchase two high efficiency heat pumps for $28,000 that will have a future salvage value of $3,000 at the end of their 15-year useful life.

The new compressor will save the company $6,500 per year in electricity costs, and the heat pumps will save $8,500 per year. The company's discount rate is 12%.

Solution: BCR (1) = PV(benefits)/PV (costs)
               PV (benefits 1) = A[P | A, i, N]
                             = $6,500/yr [P | A, 12%, 15 yr] + $2,000[P | F, 12%, 15] yr
                             = $6,500 × 6.8109 + $2,000 × 0. 1827
                             = $44,636.25
               BCR (1) = *$44,636.25* $20,000
                         = **2.23**

               BCR (2) = PV(benefits)/PV (costs)
               PV (benefits 2) = A[P | A, i, N]
                             = $8,500/yr [P | A, 12%, 15 yr] + $3,000[P | F, 12%, 15] yr
                             = $8,500 × 6.8109 + $3,000 × 0. 1827
                             = $58,440.75
               BCR (2) = *$58,440.75/$28,000*
                         = **2.09**

Therefore, since BCR(1) > BCR(2), select option 1: the new compressor.

               LCC (1) = Purchase cost - PV (benefits 1)
                       = $20,000 - $44,636.25
                       = $(24,636.25)

$$\begin{aligned} \text{LCC (2)} &= \text{Purchase cost - PV (benefits 2)} \\ &= \$28,000 - \$58,440.75 \\ &= \$(30,440.75) \end{aligned}$$

Therefore, the answer with the LCC is different. Since LCC (2) is more negative (less cost), select option 2: the two high efficiency heat pumps.

Additional Learning Point

Why the difference? While the BCR and NPV methods will provide the same accept or reject decisions on independent projects, these different methods may yield different rank orders of projects profitabilities for mutual exclusive projects. The difference is that the BCR method is a measure of how much each dollar invested earns. However, it does not take into account the overall size of the project. Therefore, to make a decision on which mutually exclusive project to select, one needs to use a NPV method, which takes into account the size (amount invested) of the project.

Problem: There are a number of energy-related problems that can be solved using the principles of economic analysis. Apply your knowledge of these economic principles to answer the following questions.

Given: a) Estimates of our use of coal have been made that say we have a 500 years' supply at our present consumption rate. How long will this supply of coal last if we increase our consumption at a rate of 7% per year? Why don't we need to know what our present consumption is to solve this problem?

               b) Some energy economists have said that it is not very important to have an extremely accurate value for the supply of a particular energy source. What can you say to support this view?

               c) A community has a 100 MW electric power plant, and their use of electricity is growing at a rate of 10% per year. When will they need a second 100 MW plant? If a new power plant costs $1 million per MW, how much money (in today's dollars) must the community spend on building new power plants over the next 35 years?

Solution::

a) Year	(yrs)	Present Use	Remaining (yrs)	Year	Present (yrs)	Use	Remaining (yrs)
1	500	1.00	500.00	27	500	6.21	420.30
2	500	1.07	498.93	28	500	6.65	413.65
3	500	1.14	497.79	30	500	7.11	406.54
4	500	1.23	496.56	31	500	7.61	398.93
5	500	1.31	495.25	32	500	8.15	390.78
6	500	1.40	493.85	33	500	812	382.07
7	500	1.50	492.35	34	500	9.33	372.74
8	500	1.61	490.74	35	500	9.98	362.76
9	500	1.72	489.02	36	500	10.68	352.09
10	500	1.84	487.18	37	500	11.42	340.66
11	500	1.97	485.22	38	500	12.22	328.44
12	500	2.10	493.11	38	500	13.08	315.36
13	500	2.25	480.86	40	500	13.99	301.36
14	500	2.41	478.45	41	500	14.97	296.39
15	500	2.58	475.87	42	500	16.02	270.37
16	500	2.76	473.11	43	500	17.14	253.22
17	500	2.95	470.16	44	500	18.34	234.88
18	500	3.16	467.00	45	500	19.63	215.25
19	500	3.38	463.62	46	500	21.00	194.25
20	500	3.62	460.00	47	500	22.47	171.79
21	500	3.87	456.13	48	500	24.05	147.73
22	500	4.14	451.99	49	500	25.73	122.00
23	500	4.43	447.56	50	500	27.53	94.47
24	500	4.74	442.82	51	500	29.46	65.01
25	500	5.07	437.75	52	500	31.52	33.50
26	500	5.43	432.32	53	500	33.73	(0.23)
26	500	5.81	426.52				

Therefore, with a present amount of coal of 500 years at the present use will only last 52 years if the use is increased by 7% a year. We do not need to know our present consumption, since we can state the consumption in terms of years.

b) New technologies will allow more efficient use of these resources. Additionally, new technologies will allow for more of these resources to be found. Furthermore, technological development will find new energy sources.

c) Therefore, one can calculate when a new power plant is needed as follows:

yr	Present peak use (MW)	yr	Present peak use(MW)	yr	Present peak use (MW)	yr	Present peak use (MW)
0	100	9	236	18	**556**	**27**	**1311**
1	**110**	10	259	**19**	612	28	1442
2	121	11	285	20	673	29	1586
3	133	**12**	314	21	740	30	1745
4	146	13	345	22	814	31	1919
5	161	14	380	23	895	32	2111
6	177	**15**	**418**	**24**	985	33	2323
7	195	16	459	**25**	1083	34	2555
8	**214**	**17**	**505**	26	1192	35	2810

Therefore, they need a new plant in year 1.

Therefore, they will need to build 28 100 MW power plants over the next 35 years. Assuming that they build the plants in 100 MW increments, a MARR of 10% and that the cash flow for building the plant all occurs in the year before they reach the next 100 MW increment (unlikely), then the present value of these plants can be calculated as follows:

yr	number of plants	cost ($million)	PV ($million)
1	1	100	90.91
8	1	100	46.65
12	1	100	31.68
15	1	100	23.94
17	1	100	19.78
19	1	100	16.35
21	1	100	13.51
22	1	100	12.28
24	1	100	10.15
25	1	100	9.23
26	1	100	8.39
27	2	200	15.26
28	1	100	6.93
29	1	100	6.30
30	2	200	11.46
31	2	200	10.42
32	2	200	9.47
33	2	200	8.61
34	2	200	7.83
35	3	300	10.68
	28		**$ 370**

Therefore, these plants will cost about $370 million in today's dollars.

Problem: How many hours per week must the gymnasium be used in order to justify the cost difference of a one-year payback?

Given: A church has a gymnasium with sixteen 500 Watt incandescent ceiling lights. An Equivalent amount of light could be produced by sixteen 250 Watt PAR (parabolic aluminized reflector) ceiling lamps. The difference in price is $10.50 per lamp, with no difference in labor. The gymnasium is used 9 months each year. Assume that the rate schedule used is that of Problem 3.8, that gymnasium lights do contribute to the peak demand (which averages 400 kW), and that the church consumes enough electricity that much of the bill comes from the lowest cost block in the table.

Solution: <u>Customer charge</u>: $8.22/mo

     <u>Energy charge</u> $0.1230/KWh   on-peak
                   $0.0489/kWh   off-peak

This rate charges less for electricity used during off-peak hours—about 80% of the hours in a year—than it does for electricity used during on-peak hours.

   AC = (Off-peak percentage of energy use) (Off-peak energy cost) + (1 - off-peak percentage of energy use)(On-peak energy cost)

Therefore, the average cost per kWh with the above schedule is:

   AC = (80%)($0.0489/kWh)
          + (1 - 80%)($0.123/kWh)
       = *$0.06372/kWh*

The demand reduction (DR) from the retrofit can be calculated as follows:

   DR = N × (Do - Dnew)

where,

N = Number of lamps, 16 lamps

Do = Initial demand per lamp, 500 W/lamp

$Dnew$ = Demand of the PARs per lamp, 250 W/lamp

Therefore,

DR = 16 lamps × (500 W/lamp - 250 W/lamp)

       = 4,000 W

       = 4 kW

The implementation cost (IC) can be calculated as follows:

IC = Cost premium × N

       = $10.50 × 16 lamps

       = $168.00

SPP = IC/CS

where,

CS = Cost savings

Therefore

CS = IC/SPP = $168.00/yr = $168.00/yr

The number of weeks (Nw) the gym is used in the world can be estimated as follows:

Nw = 52 wks/yr × 9 months/12 months = 39 wks/yr

The number of hours a week (h) the lights must operate can be calculated as follows:

CS = DR × h × 39 wks/yr × AC

Therefore,

h = CS/(DR × 39 wks/yr × AC)

      = $168.00/(4 kW × 39 wks/yr × $0.06372/kWh)

      = 16.9 h/wk

Problem: Find the equivalent present worth and IRR of the following
              6-year project:

Given: Use the depreciation schedule in Table 4-6
              purchase and installation cost: $100,000
              annual maintenance cost: $10,000
              annual energy cost savings: $45,000
              salvage value: $20,000
              MARR: 12%
              Tax rate: 34%
              equipment life: 5 years for depreciation purposes

Depreciation Table: MACRS Percentages for 3-, 5-, and 7-year Property

Year	3-year	5-year	7-year
1	33.33%	20.00%	14.29%
2	44.45%	32.00%	24.49%
3	14.81%	19.20%	17.49%
4	7.41%	11.52%	12.49%
5		11.52%	8.93%
6		5.76%	8.92%
7			8.93%
8			4.46%
	100%	100%	100%

Solution: Assuming end of year convention

Year	Before tax cash flow	Depreciation	Taxes	After tax cash flow	PV
0	$(100,000)			$(100,000)	$(100,000)
1	$35,000	$20,000	$5,100	$29,900	$26,696
2	$35,000	$32,000	$1,020	$33,980	$27,089
3	$35,000	$19,200	$5,372	$29,628	$21,089
4	$35,000	$11,520	$7,983	$27,017	$17,170
5	$35,000	$11,520	$7,983	$27,017	$15,330
6	$55,000	$5,760	$16,742	$38,258	$19,383
				NPV:	*$26,756*
				IRR:	20.9%

The MARR that drives the present value to zero is 20.9%, which is the
IRR or ROR.

Problem: Calculate the constasnt dollar, after tax ROR or IRR for Problem 4-7 if the inflation rate is 6%.

Given: Use the depreciation schedule in Table 4-1
            purchase and installation cost: $100,000
            annual maintenance cost: $10,000
            annual energy cost savings: $45,000
            salvage value: $20,000
            MARR: 12%
            Tax rate: 34%
            Equipment life: 5 years for depreciation purposes
            Inflation rate   6%

Solution: Assuming end of year convention

Year	Before tax cash flow	Depreciation	Taxes	After tax cash flow	PV
0	$(100,000)			$(100,000)	$(100,000)
1	$37,100	$20,000	$5,814	$31,286	$33,125
2	$39,326	$32,000	$2,491	$36,835	$31,350
3	$41,686	$19,200	$7,645	$34,040	$29,671
4	$44,187	$11,520	$11,107	$33,080	$28,081
5	$46,838	$11,520	$12,008	$34,830	$26,577
6	$69,648	$5,760	$21,722	$47,926	$35,286
				NPV:	*$84,091*
				IRR:	26.5%

The MARR that drives the present value to zero is 26.5%, which is the *IRR* or ROR.

Problem: What is the constant dollar, after-tax **ROR or IRR** for this project?

Find the equivalent constant dollar after-tax present worth of the following 6-year project using the depreciation schedule in Table 4-6:

Given:

Purchase and installation cost	$100,000
Annual maintenance (AM)	$10,000/yr
Maintenance cost inflation	5%/yr
Annual energy savings (ES)	$45,000
ES growth	8%/yr
Salvage value (SV)	$20,000
Salvage value growth	6%/yr
Consumer price index (CPI) growth	6%/yr
MARR in constant dollars	12%/yr
Tax rate	34%/yr
Depreciation life (N)	5 yrs

Solution:

Year	PV	Constant Dollar	After tax cash flow	Taxes	Taxable income	Depreciation	Cash flow	AM	ES	SV
0	$(100,000)	$(100,000)	$(100,000)							
1	$25,185	$28,208	29,900	5,100	15,000	20,000	35,000	10,000	45,000	
2	$25,560	$32,063	36,026	2,074	6,100	32,000	38,100	10,500	48,600	
3	$20,256	$28,458	33,894	7,569	22,263	19,200	41,463	11,025	52,488	
4	$16,959	$26,686	33,690	11,421	33,591	11,520	45,111	11,576	56,687	
5	$15,392	$27,126	36,301	12,766	37,547	11,520	49,067	12,155	61,222	
6	$19,964	$39,406	55,898	25,829	75,967	5,760	81,727	12,763	66,120	28,370

$23,317

The MARR that drives the present value to zero is 19.69%, which is the *IRR* or *ROR*

Chapter 5

Electrical Systems

Problem

5.1 PF = 1.0 or 100% (since it is an AC resistive load)
P_{IN} = V x A x PF = 120 x 8 x 1 = 960W or 0.96kW

5.2 The larger wire has a lower resistance, and the I^2R loss in the distribution system will be lower. This reduces their cost per kWh.

5.3 Single phase AC load
$$P_{IN} = V \text{ x } A \text{ x } PF \text{ W}$$
$$= kV \text{ x } A \text{ x } PF \text{ kW}$$
$$kW_{IN} = (0.24)\ (20)\ (0.8)$$
$$= 3.84 \text{ kW}$$

5.4 Electrical load is 200 kVA and 100 kW. Find PF and kVAR

$$PF = \frac{kW}{kVA} = \frac{100}{200} = .5 \text{ or } 50\%$$

$$kVAR^2 = kVA^2 - kW^2 = 200^2 - 100^2$$
$$= 30{,}000$$
$$kVAR = \sqrt{30{,}000} = 173.2 \text{ kVAR}$$

5.5 Inductive kVAR creates the electromagnetic fields that produce the torque to start and tun the motor. kVAR does not use energy. The kW the motor draws provides the real work.

5.6 Three phase AC industion motor with 460 volts, 50 amperes, and a power factor of 72%.

What is the kW input?

$$kW_{IN} = \sqrt{3} \times kV \times A \times PF$$
$$= 1.732 \times .46 \times 50 \times 0.72$$
$$= 28.68 \text{ kW}$$

5.7   Motor is 60 HP, efficiency = 91.5%, and full load. Find kW input.

$$kW_{IN} = \frac{60 \text{ HP}}{} \left| \frac{0.746 \text{kW}}{\text{HP}} \right| \frac{LF = 1}{\text{EFF} = .915} = 48.92 \text{ kW}$$

5.8   460 volt motor, 40 amperes, and 28 kW input. Find the PF.

$$kW_{IN} = \sqrt{3} \times kV \times A \times PF$$
$$28 \text{ kW} = 1.732 \times .46 \times 40 \times PF$$
$$= 31.87 \times PF$$

$$PF = \frac{28}{31.87} = \frac{0.879}{} = \frac{87.9\%}{}$$

5.9   AC induction motor draws 150 kW, and has a PF of 85%. Find the kVARs of capacitance to put on the motor to increase the PF to 90%.

kVAR = 150 x PF table number to go from 85 to 90%
      Table number = 0.136
kVAR = 150 x 0.136 = 20.4 kVAR

5.10

a) Monthly load factor

$$MELF = \frac{350,000 \text{ kWh}}{600 \text{ kW per}} \times 720 \text{ hours} = 81\%$$

b) Average cost per kWh
    Energy cost   = 350,000 kWh x $0.12/kWh
                  = $42,000

$$kW_{IN} = \cfrac{600\ kW\ \Big|\quad \$8\quad \Big|\ 1\ Mo}{\Big|\quad kW\ Mo\quad \Big|}$$

$$= \$4,800$$

Total cost $= \$42,000 + \$4,800$

$$= \$46,800$$

Average cost

per kWh $= \$46,800 / 350,000\ kWh$

$$= \$0.1337 / kWh$$

c) If the monthly load factor had been 25%, for othe same kWh
find the peak kW for the month MELF = 0.25

$$= \cfrac{350,000\ kWh}{kW_{PEAK} \times 720\ h}$$

$$kW_{PEAK} = \cfrac{350,000\ kWh}{0.25 \times 720} = 1944.4\ kW$$

Find average kWh cost.

Energy cost still $42,000

Demand cost $= 1044.4\ kW \times \$8/kW = \$15,555.20$

Total cost $= \$15,555.20$

$$\underline{\$42,000.00}$$

$$\$57,555,20$$

Average cost per kWh $= \$57,555,20 / 350,000\ kWh$

$$= \$0.1644 / kWh$$

Chapter 6

Lighting

6.1

Problem: *How* much can you save by installing a photocell?
What is the payback period of this investment?

Given: When performing an energy survey, you find twelve two-lamp F40T12 security lighting fixtures turned on during daylight hours (averaging 12 hours/day). The lamps draw 40 Watts each, the ballasts draw 12 Watts each, and the lights are currently left on

24 hours per day.

Energy cost:	$0.055	/kWh
Power cost:	$7	/kW
Lamps:	$1	/lamp
Photocell (installed):	$85	/cell

Solution: Assuming one photocell can control all 12 fixtures

There will probably be demand savings, since the lights will be turned off during the day. It is probable that their peak demand occurs during the day. Therefore, the demand reduction (DR) can be calculated as follows:

$$DR = Nf \times N1 \times P1 + Nf \times Pb$$

where,

Nf = Number of fixtures, 12 fixtures
$N1$ = Number of lamps per fixture, 2 lamps/fixture
$P1$ = Power use of lamps, 40 W/lamp
Pb = Power use of ballasts, 12 W/fixture

Therefore,

$$DR = 12 \text{ fixtures} \times 2 \text{ lamps/fixture} \times 40 \text{ W/lamp} +$$
$$12 \text{ fixtures} \times 12 \text{ W/fixture}$$
$$= 1,104 \text{ W}$$
$$= 1.104 \text{ kW}$$

Therefore, the energy savings (ES) can be calculated as follows:
$$ES = DR \times 12 \text{ hr/day} \times 365 \text{ days/yr}$$
$$= 1.10 \ 4kW \times 12 \text{ hr/day} \times 365 \text{ days/yr}$$
$$= 4,835.52 \text{ kWh/yr}$$

Therefore, the cost savings (CS) can be calculated as follows:
$$CS = ES \times \$0,055/kWh +$$
$$DR \times \$7/kW \times 12 \text{ mo/yr}$$
$$= \mathbf{\$358.69/yr}$$

$$\mathbf{\textit{SPP}} = IC/CS$$
$$= \$85/\$358.69/yr$$
$$= 0.24 \text{ years}$$
$$= \mathbf{\textit{2.84 months}}$$

6.2

Problem: What is the simple payback period (SPP) and what is the return on investment for each alternative?

Given: You count 120 four-lamp F40T12 troffers that contain 34-Watt lamps and two ballasts. How much can you save by installing:
a. 3-F40T10 lamps at $15/fixture?
b. 3-F32T8 lamps and an electronic ballast at $40/fixture?
Assume the same energy costs as in problem 5. 1.

Solution:

Energy cost:	$0.055 /kWh
Power cost:	$7 /kW
Number of fixtures:	120 fixtures
Present power use
per lamp (Pp)	156.4 W/fixture including ballast
Power use per fixture
for option a. (Pa)	138 W/fixture including ballast
Power use per fixture
for option b. (Pb)	91.2 W/fixture including ballast
Implementation cost
for option b. (ICb)	$15 /fixture
Implementation cost
for option a. (ICa)	$40 /fixture
Assuming that the
lights are used	8,760 hrs/yr
Assuming that the
life of the fixtures is	7 yrs

Option	Demand Reduction (kW)	Energy Savings (kWh/yr)	Cost Savings ($/yr)	IC ($)	SPP (yrs)	IRR (%)
a.	2.21	19,342	1,249	1,800	*1.44*	*67.5*
b	7.82	68,538	4,427	4,800	*1.08*	*91.2*

6.3 Problem: How much can you save by replacing the two 20-Watt bulbs with a 7-Watt CFL?

Given: You see 25 exit signs with two 20-Watt incandescent lamps each. The 20-Watt incandescent lamps have a 2,500-hour life span and costs $3 each. The 7-Watt CFLs have a 12,000-hour life span and cost $5 each and require the use of a $15 retrofit kit. Assume the same energy costs given in Problem 5-1.

Solution:

Energy cost:	$0.055	/kWh
Power cost:	$7	/kW
Number of fixtures:	25	fixtures
Present power use per lamp (Pp)	40	W/fixture
Power use per fixture of retrofit (Pr)	7	W/fixture	including ballast
Present life of lamps (Lp)	2,500	hours/lamp
Life of retrofit lamps (Lr)	12,000	hours/lamp
Present lamp cost (cp)	$3	/lamp
Retrofit lamp cost (cr)	$5	/lamp
Assuming that the lights are used	8,760	hrs/yr

Assuming that the labor needed to replace the lamp is included in lamp replacement cost

Option	Demand Reduction (kW)	Energy Savings (kWh/yr)	Energy and Demand Cost Savings ($/yr)	Annual Lamp Replacement Costs ($/yr)	Lamp Replacement Cost Savings ($/yr)	Total Annual Cost Savings ($/yr)
Present	—	—	—	525.60	—	—
Retrofit	0.825	7,227	466.79	365.00	160.00	627.39

The annual lamp replacement costs (LRC) can be calculated as follows:

LRC = Number of lamps per fixture × Number of fixtures × Replacement lamp cost × Annual lamp use/Lamp life

6.4

Problem: How can this problem be solved, and how much money can you save in the process?

Given: An old train station is converted to a community college center, and a train still passes by in the middle of the night. There are 82 75-Watt A19 lamps in surface-mounted wall fixtures surrounding the building, and they are turned on about 12 hours per day. The lamps cost $0.40 each and last for about one week before failure. Assume electricity costs 8 cents per kWh.

Solution: According to references such as *The Lighting Handbook*, a replacement for a 75-Watt incandescent lamp is an 18-W compact fluorescent lamp (CFL). CFL last longer than incandescent. Additionally, according to Table 5-10 this will have energy savings $34.20 over the life of the 18-W CFL. Additionally, according to Table 5-5, one expects an 18-W CFL to last 10,000 hours. Since the lamps are on half the time (12-hours per day), we expect the CFL to last 2 years. Additionally, since at the present time, the train station personnel replace the lamps about once a week at a cost of $0.40 per lamp or $20.80 per year (52 weeks × $0.40/wk). Additionally, we expect 18-W CFL to cost about $20 per lamp. Therefore, the cost of the replacement lamps cancel. Therefore, we can calculate the energy cost savings as follows:

$$\text{ECS} = \$34.20/\text{lamp}/2 \text{ years} \times 82 \text{ lamps}$$
$$= \$1,402.20/\text{yr}$$

Additionally, one would expect a labor cost savings. Assuming that the burdened labor cost is $10 per hour and the maintenance crew spends 2 man-hr/wk to replace the lamps. Therefore, the labor savings (LS) can be calculated as follows:

$$\text{LS} = \$1 0/\text{man-hr} \times 2 \text{ man-hr/wk} \times 52 \text{ wk/yr}$$
$$= \$1,040.00/\text{yr}$$

Therefore, we estimate the total annual cost savings (CS) as follows:

$$\text{CS} = \text{Lamp replacement savings} + \text{ECS} + \text{LS}$$
$$= 0 + \$1,402.20/\text{yr} + 1,040/\text{yr}$$
$$= \boldsymbol{\$2,442.20/yr}$$

6.5

Problem: How much can you save by replacing these fixtures with 70-Watt HPS cutoff luminaires?

Given: During a lighting survey you discover thirty-six 250-Watt mercury vapor cobrahead streetlights operating 4,300 hours per year on photocells.
There is no demand charge, and energy costs $0.055 per kWh.

Solution:

Energy cost: $0.055 /kWh
Number of fixtures: 36 fixtures
Present power use per lamp (Pp)   300 W/fixture including ballast
Power use per fixture of retrofit (Pr)   84 W/fixture including ballast
Assume both lamp lives are about the same
Assume both lamp costs are about the same
Lamp use   4,300 hrs/yr

Option	Demand Reduction (kW)	Energy Savings (kWh/yr)	Energy and Demand Cost Savings ($/yr)
Present	—	—	—
Retrofit	7.776	33,437	**1,839.02**

6.6

Problem: What is the savings from retrofitting the facility with 250-
Watt high pressure sodium (HPS) downlights?
What will happen to the lighting levels?

Given: You find a factory floor that is illuminated by eighty-four
400-Watt mercury vapor downlights. This facility operates
two shifts per day for a total of 18 hours, five days per
week.
Assume that the lights are contributing to the facility's
peak demand, and the rates given in Problem 5-1 apply.

Solution:

Energy cost:	$0.055 /kWh
Power cost:	$ 7/kW
Number of fixtures:	84 fixtures
Present power use per lamp (Pp)	480 W/fixture including ballast
Power use per fixture of retrofit (pr)	300 W/fixture including ballast
Assume lamp lives are about the same	
Assume lamp costs are about the same	
Assuming that the lights are used	4,680 hrs/yr

Option	Demand Reduction (kW)	Energy Savings (kWh/yr)	Energy and Demand Cost Savings ($/yr)
Retrofit	15.120	70,762	5,161.97

*One would expect the lighting levels to be about the same
to a little higher.* Check the manufacturer's data for an exact
comparison.

6.7

Problem: What will happen to the lighting levels throughout the space and directly under the fixtures? Will this retrofit be cost-effective?
What is your recommendation?

Given: An office complex has average ambient lighting levels of 27 foot-candles with four-lamp F40T12 40-Watt 2' × 4' recessed troffers. They receive a bid to convert each fixture to two centered F32T8 lamps with a specular reflector designed for the fixture and an electronic ballast with a ballast factor of 1. 1 for $39 per fixture. This lighting is used on-peak, and electric costs are $6.50 per kW and $0.05 per kWh.

Solution: *One would expect the overall lighting levels to decrease, while the reflectors should reduce this reduction by concentrating the light to the areas below the lights.*

Energy cost:	$0.05 /kWh
Power cost:	$6.50 /kW
Number of fixtures:	1 fixtures
Present power use per lamp (Pp)	184.0 W/fixture including ballast
Power use per fixture for retrofit (Pr)	70.4 W/fixture including ballast
Implementation cost for retrofit (IC)	$39 /fixture
Assuming that the lights are used	8,760 hrs/yr

Option	Demand Reduction (kW)	Energy Savings (kWh/yr)	Savings ($/yr)	Cost IC ($)	SPP (yrs)
retrofit	0.11	995	59	39	**0.67**

Since this retrofit would payback in less than a year, this would be a good project.

6.8

Problem: What is your advice?

Given: An exterior loading dock in Chicago uses F40T12 40-Watt lamps in enclosed fixtures. They are considering a move to use 34-Watt lamps.

Solution:

Energy cost:	$0.055	/kWh
Power cost:	$7.00	/kW
Number of fixtures:	1	fixtures
Present power use per lamp (Pp)	46.0	W/fixture including ballast
Power use per fixture for retrofit (Pr)	39.1	W/fixture including ballast
Implementation cost for retrofit (IC)	$1	/fixture
Assuming that the lights are used	8,760	hrs/yr

Option	Demand Reduction (kW)	Energy Savings (kWh/yr)	Cost Savings ($/yr)	IC ($)	**SPP (yrs)**
retrofit	0.01	60	4	1	**0.26**

At times, you may need to collect more information from your client based on your energy bill analysis or your audit about hours of operations. When there is variability in factors, you may need to perform sensitivity analysis. One good approach is to give a best-case, worst-case, and most likely cause.

If we assume that the electric costs are the same as in Problem 5-1, the lights are on all the time, the cost of the 34-Watt lamps is $1 more per, lamp, and the lighting levels are higher than needed (the 34-Watt lamps will produce a little less light), then this project looks good since its payback is less than a year.

Therefore, my advice would be to replace the 40-Watt lamps with 34-Watt lamps the next time they perform a group lamp replacement.

6.9

Problem: How would you recommend they proceed with lighting changes? What will be the savings if they have a cost of 6 cents per kWh?

Given: A turn-of-the-century power generating station uses 1500-Watt incandescent lamps in pendant mounted fixtures to achieve lighting levels of about 18 foot-candles in an instrument room. They plan on installing a dropped ceiling with a 2' × 4' grid.

Solution: There exists many strategies that could work depending on other conditions such as the need for light at various work surfaces and the height of the ceiling. One strategy that may work is replacing each 1,500-Watt lamp with a four-lamp F32T8 fixture. This would work if the lighting level remains within acceptable levels, which could depend on how far the drop ceiling lowers the lamps towards the working surface. Perhaps, this strategy could be used in combination with task lighting. Assuming that the one four-lamp F32T8 fixture and a 18-Watt CFL for each 1,500-Watt lamp provides an acceptable lighting level, then the cost savings (CS) from this retrofit can be calculated as follows:

Energy cost:	$0.06	/kWh
Number of fixtures:	1	fixtures
Present power use per lamp (Pp)	1,500	W/fixture
Power use per fixture for F32T8 (Pf)	121.6	W/fixture including ballast
Power use per fixture for CFL (Pc)	18	W/fixture including ballast
Assuming that the lights are used	8,760	hrs/yr

Option	Demand Reduction (kW)	Energy Savings (kWh/yr)	Cost Savings ($/Yr)	
retrofit	1.36	11,917	**715**	per fixture

6.10

Problem: What would be the life-cycle savings of using 13-Watt CFL in the same fixtures?

Given: A meat-packing facility uses 100-Watt A19 lamps in jarlights next to the entrance doors. These lamps cost $0.50 each and last 750 hours. The CFLs cost $15 each, and last 12,000 hours. The lights are used on-peak, and the electricity costs 8 cents per kWh.

Solution:

Energy cost:	$0.08 /kWh
Number of fixtures:	1 fixtures
Present power use per lamp (Pp)	100 W/fixture
Power use per fixture of retrofit (Pr)	13 W/fixture including ballast
Present life of lamps (Lp)	750 hours/lamp
Life of retrofit lamps (Lr)	12,000 hours/lamp
Present lamp cost (cp)	$0.50 /lamp
Retrofit lamp cost (Cr)	$15 /lamp
Assuming that the lights are used	8,760 hrs/yr
Assuming the MARR is	15%

Assuming that the labor needed to replace the lamp is included in lamp replacement cost

Option	Demand (kW)	Energy Use (kWh/yr)	Energy and Demand Cost in one life ($/yr)	Annual Lamp Replace-ment- Costs ($/yr)	PV over 12,000 hours	Total Cost Savings over 12,000 hours ($)
100-W inc	0.100	876	70.08	5.84	93.81	—
F-13-W CFL	0.013	114	9.11	10.95	24.79	69.02

6.11

Problem: What problems can you anticipate from the light trespass off the lot? How would you recommend improving the lighting? How much can you save with a better lighting source and design?

Given: A retail shop uses a 1,000-Watt mercury vapor floodlight on the corner of the building to illuminate the parking lot. Some of this light shines out into the roadway. Use the electric costs from Problem 5-7, and assume the light does not contribute to the shop's peak load.

Solution: The problems include possible liability and wasting energy by lighting an area that does not need light. One could improve the lighting design by properly aiming the light (similar to Figure 5-7) and using a more efficient light source: Table 5-10 recommends using a 880-Watt High Pressure Sodium (HPS).

Energy cost:	$0.05	/kWh
Number of fixtures:	1	fixtures
Present power use per lamp (Pp)	1,200	W/fixture including ballast
Power use per fixture for retrofit (Pr)	1,056	W/fixture including ballast
Assuming that the lights are used	4,380	hrs/yr

Option	Demand Reduction (kW)	Energy Savings (kWh/yr)	Cost Savings ($/yr)
retrofit	0.144	631	**$31.54**

6.12

Problem: What are the energy, power, and relamping savings from using two 250-Watt HPS floodlights? What will happen to the lighting levels?

Given: A commercial pool uses four 300-Watt quartz-halogen floodlights. The lights do contribute to the facility's peak load, and the electric rates are those of Problem 5-7.

Solution:

Energy cost:	$0.05 /kWh
Demand cost:	$6.50 /kW
Present number of fixtures:	4 fixtures
Proposed number of fixtures:	2 fixtures
Present power use per lamp (Pp)	300 W/fixture including ballast
Power use per fixture for retrofit (Pr)	300 W/fixture including ballast
Assuming that the lights are used	4,380 hrs/yr

Option	Demand Reduction (kW)	Energy Savings (kWh/yr)	Cost Savings ($/yr)
retrofit	**0.6**	**2,628**	**$178.20**

The lighting level will be increased.

6.13

Problem: What is the solution?

Given: You notice that the exterior lighting around a manufactur-
               ing plant is frequently left on during the day. You are told
               that this is due to safety-related issues. Timers or failed
               photocells would not provide lighting during dark overcast
               days.

Solution: The photocell sensitivity could be set to provide light even
               during dark overcast days. Another solution could be to
               provide a mixture of low sensitivity photocells and more
               sensitive photocells. in this way a proportion of the lights
               would come on during overcast days and the rest would
               only come on during the night or extremely dark days.

6.14

Problem: How can you solve these problems?

Given: A manufacturing facility uses F96T12HO lamps to illuminate
              the production area. Lamps are replaced as they burn out.
              These fixtures are about 15 years old and seem to have a
              high rate of lamp and ballast failure.

Solution: One could retrofit the system with a newer lighting system.
              For example, a system using T8 lamps with electronic bal-
              lasts seems appropriate. Additionally, they should imple-
              ment a group relamping program, which would eliminate
              the need to replace lamps one-by-one as they fail. Thereby,
              these two recommendations would not only save energy, but
              it would also save labor costs.

Chapter 7

Motors and Drives

Problems

1. Most AC induction motors have no load speeds of 3600 or 1800
    RPM. So if the full load speed of the motor is 1765 RPM, it is an
    1800 RPM motor.

2. A 30 HP AC motor draws 25 kW at full load. Find efficiency.

    $$\text{Efficiency} = \frac{\text{output}}{\text{input}} = \frac{30 \times .746 \text{ kW}}{25 \text{ kW}}$$

    $$= 89.5\% = \frac{22.4}{25}$$

3. 30 HP motor drives a 15 HP pump. Find the motor load factor.

    $$\text{Load Factor} = \frac{15 \text{ HP}}{30 \text{ HP}} = 50\%$$

4. 30 HP 3 phase motor, 460 volts, FLA = 37 amps. Find the power
    factor kW_{IN} from Q3 = 25 kW.

    $$kW \sqrt{3} = \times \text{ kV} \times \text{ A} \times \text{ PF}$$
    $$25 = (1.732) \times 0.46 \times 37 \times \text{PF}$$
    $$25 = 29.48 \times \text{PF}$$

    $$\text{PF} = \frac{25}{29.48} = 0.85$$

73

5. 40 HP AC induction motor, full load speed = 1760 RPM
    Running speed = 1773 RPM

$$\text{Load factor} = \frac{\text{No Load RPM} - \text{Running RPM}}{\text{No Load RPM} - \text{Full Load RPM}}$$

$$= \frac{1800 - 1773}{1800 - 1760} = \frac{27}{40}$$

$$= 0.675 = 67.5\%$$

6. 100 HP induction motor
    LF = 0.70 EFF = 92.3%

$$kW_{IN} = \frac{100 \text{ HP}}{} \left| \frac{0.746 \text{ kW}}{\text{HP}} \right| \frac{\text{LF} = 0.70}{\text{EFF} = .923}$$

    kW_{IN} = 56.58 kW

7. Three phase AC induction motor 460 volts I = A = 80 amps
    PF = 85% Find kW_{INPUT}

$$
\begin{aligned}
kW_{INPUT} &= \sqrt{3} \times kV \times A \times PF \\
&= (1.732) \times (0.46) \times (80) \times (0.85) \\
&= 54.18 \text{ kW}
\end{aligned}
$$

8. 100 kW AC induction motor PF = 0.75
    How many kVAR should be added to have a new PF = .85?
    kVAR = 100 kW (table number) 75% —> 85%
        = 100 (0.262)
        = 26.2 kVAR

9. 40 HP AC induction motor
    LF = 75%   EFF = 89.3%
    New 30 HP   LF = 100%   EFF = 93.6%
    How many kW of savings?

$$kW_{savings} = \left[\frac{40 \text{ HP}}{} \left| \frac{.746 \text{ kW}}{\text{HP}} \right| \frac{.75}{.893} \right]$$

$$-\left[\frac{30\ \text{HP}}{}\left|\frac{.746\ \text{kW}}{\text{HP}}\right|\frac{1.0}{.936}\right]$$

$= 25.062 - 23.91$
$= 1.152\ \text{kW}$

10. 10 HP centrifugal fan 2000 CFM. On a cooler day it supplies 1500 CFM by a VFD. How many HP is needed now?

$$\frac{\text{HP}_{\text{NEW}}}{\text{HP}_{\text{OLD}}}=\left(\frac{\text{CFM}_{\text{NEW}}}{\text{CFM}_{\text{OLD}}}\right)^{3}$$

$$\text{HP}_{\text{NEW}}=\text{HP}_{\text{OLD}}\left(\frac{1500}{2000}\right)^{3}$$

$$\qquad\qquad 10\qquad\quad .75$$

$= 10\ (.75)^3$
$= 10\ (.422)$
$= 4.22\ \text{HP}$

Chapter 8

Heating, Ventilating, and Air Conditioning

8.1

Problem: Estimate the heating load.

Given: The heating load of a facility is due to a work force of 22 people including 6 overhead personnel, primarily sitting during the day; 4 maintenance personnel and supervisors; and 12 people doing heavy labor. Assume everyone works the same 8-hour day.

Solution: Assuming all the people are males, the heat load (q) can be estimated as follows:

$$q = [Ns \times qs + Nn \times qn + Nh \times qh] \times h$$

where,

Ns = Number of people seated, 6 people

qs = Heat gain from seated people, 400 Btu/h/person

Ns = Number of people doing light machine work, 4 people

qs = Heat gain from people doing light machine work, 1,040 Btu/h/person

Ns = Number of people doing heavy work, 12 people

qs = Heat gain from people doing heavy work, 1,600 Btu/h/person

h = Number of working hours, 8 hrs/day

Therefore,

q = [6 people × 400 Btu/h/person + 4 people × 1,040 Btu/hr/person + 12 people × 1,600 Btu/h/person] × 8 hrs/day

= *206,080 Btu/day*

= *25,760 Btu/h*

8.2

Problem: How many kW will this load contribute to the electrical peak
              if the peak usually occurs during the working day?

Given: The HVAC system that removes the heat in Problem 6.1 has
              a COP of 2.0 and runs continuously. Assume that the motors
              in the HVAC system are outside the conditioned area and
              do not contribute to the cooling load.

Solution:

$$
\begin{aligned}
EER &= COP \times 3.412 \text{ Btu/Wh} \\
    &= 2 \times 3.412 \text{ Btu/Wh} \\
    &= 6.824 \text{ Btu/Wh} \\
W &= \text{Btu/h cooling/(EER)} \\
  &= 25{,}760 \text{ Btu/h}/6.824 \text{ Btu/Wh} \\
  &= 3{,}774.9 \text{ W} \\
  &= \textit{3.77 kW}
\end{aligned}
$$

8.3

Problem: Answer Problem 6.2 with the following assumptions:

Given: 8 of the 12 people doing heavy labor and 2 foremen-maintenance personnel come to work when the others are leaving and that 3,000 W of extra lighting are required for the night shift.

Solution: Assuming all the people are males, the heat load (q) can be estimated as follows:

$$q = [\text{Lighting load} \times 3{,}412 \text{ Btu/kWh} + \text{Nn} \times qn + \text{Nh} \times qh] \times h$$

where,

Ns = Number of people doing light machine work, 2 people

qs = Heat gain from people doing light machine work, 1,040 Btu/h/person

Ns = Number of people doing heavy work, 8 people

= Heat gain from people doing heavy work, 1,600

Btu/h/person

h = Number of working hours, 8 hrs/day

Therefore,

$$q = [3 \text{ kW} \times 3{,}412 \text{ Btu/kWh} + 2 \text{ people} \times 1{,}040 \text{ Btu/hr/person} + 8 \text{ people} \times 1{,}600 \text{ Btu/h/person}] \times 8 \text{ hrs/day}$$
$$= 200{,}928 \text{ Btu/day}$$
$$= 25{,}116 \text{ Btu/h}$$

$$W = \text{Btu/h cooling/(EER)}$$
$$= 25{,}116 \text{ Btu/h } 6.824 \text{ Btu/Wh}$$
$$= 3{,}680.5 \text{ W}$$
$$= \textbf{3.68 kW}$$

8.4

Problem: A 20-ton air conditioning unit has an EER of 9.3. What is
the COP of that unit? What is the kW/ton for that unit?
How many kW of input power does that unit draw when
it operates at full load mechanical cooling?

Solution: A 20-ton AC unit has an EER of 9.3 What is its COP?

Solution

COP = EER/3.412 = 2.73

Find the kWIN for this unit.

$$\frac{kW_{IN}}{ton} = \frac{12}{EER} = \frac{12}{9.3} = 1.29 \; \frac{kW}{ton}$$

8.5

Problem: If Crown Jewels buys a new motor, which one of these incentives should they ask for?

Given: Florida Electric Company offers financial incentives for large customers to replace their old electric motors with new, high efficiency motors. Crown Jewels Corporation, a large customer of FEC, has a 20-year-old 100-hp motor that they think is on its last legs, and they are considering replacing it. Their old motor is 91% efficient, and the new motor would be 95% efficient. FEC offers two different choices for incentives: either $6/hp (for the size motor considered) incentive or; a $150/kW (kW saved) incentive.

Solution: Assume a load factor of 0.6

$$\text{P saved} = 100 \text{ hp} \times 0.746 \text{ kW/hp} \times$$
$$0.6 \times [(1/0.91) - (1/0.95)]$$
$$= 2.07 \text{ kW}$$
$$\$ \text{ incentive for kW saved} = \$311$$

$$\$ \text{ incentive for size of motor} = \$600$$

Therefore, they should ask for the $6/hp (size of motor) incentive.

8.6

Problem: What is the implied efficiency of a motor if we say its load is 1 kW per hp?

What is the implied COP of an air conditioner that has a load of 1 kW per ton?

Given: Our "rules of thumb" for the load of a motor and air conditioner have implicit assumptions on their efficiencies.

Solution:

$$\text{eff} = 0.746 \text{ kW/hp} \times 1 \text{ hp/kW}$$
$$= 74.6\%$$
$$\text{COP} = 1 \text{ ton/kW} \times 12,000 \text{ Btu/tonh} \times \text{kWh/3,412 Btu}$$
$$= 3.52$$

8.7

Problem: A window air conditioner is rated at 5000 Btu/hour, 115 volts, 7.5 amps. Assuming that the power factor has been corrected to 100%, what is its SEER? How many kWh are used if the unit runs 2000 hours each year at full load? What is the annual cost of operation if electric energy costs 7.5 cents per kWh? How many kWh would be saved if the unit had an SEER of 9.1? How much money would be saved? Compute three economic performance measures to show whether this more efficient unit is a cost-effective investment. The low efficiency unit costs $200, the higher efficiency unit $250, and each unit lasts ten years. Use a MARR of 15%.

a) What is the air conditioner's SEER?
b) How many kWh are used if the unit runs 2,000 hours each year?
c) What is the annual cost of operation if electric energy costs 7.5 cents per kWh?
d) How many kWh would be saved if the unit had an SEER of 9.1?
e) How much money would be saved?
f) Show whether this more efficient unit is a cost-effective investment.

Given: A window air conditioner is rated at 5, 000 Btu/hr, 115 volts, 7.5 amps. Assume that the power factor has been corrected to 100%. The low efficiency unit costs $200, the higher efficiency unit $250, and each unit lasts ten years.

Solution: a) EER   = Btu/h [W
           = 5,000 Btu/hr/(115 v × 7.5 a)
           = 5.80 Btu/Wh
        b) kWh/yr = Running hours per year × Btu/h cooling/(EER)
           = 2,000 hr/yr × 5,000 Btu/hr/5.80 Btu/Wh
           = 1,725,000 Wh/yr
              1, 725 kWh/yr
        c) Annual cost = 1,725 kWh/yr × $0.075/kWh
           = $129.38/yr

d)

kWh/yr (9. 1)	=	Running hours per year × Btu/h cooling/(EER)
	=	2,000 hr/yr × 5,000 Btu/hr/9.1 Btu/Wh
	=	1,098,901 Wh/yr
	=	1,099 kWh/yr

kWh/yr
   (saving)   = kWh/yr (5.8) - kWh/yr (9. 1)
                  = 1,725 kWh/yr – 1,099 kWh/yr
                  = 626 kWh/yr

e)   CS   = kWh/yr (saving) × \$0.075/kWh
                  = 626 kWh/yr × \$0.075/kWh
                  = \$46.96/yr

f)   SPP   = Cost premium/CS
                  = (\$250 - \$200)/\$46.96/yr
                  = 1.06 yrs

    IRR   = 93.79%
Assuming a MARR of 15%:
    NPV   = \$185.68
    BCR   = 4.71

This looks like a good project.

8.8

Problem: How many CFM of sensible heat in Btu/h is needed to heat 1000 CFM of 35 degree F air to 95 degree F air?

Solution: Use the ASHRAE sensible heat formula.

$$q = CFM \times 1.08 \times \Delta T \quad Btu/hr$$
$$= 1000 \times 1.08 \times (95{-}35) = 64,800 \; Btu/h$$

8.9

Problem: An air conditioner removes 120,000 Btu/hr of heat, and has an input power of 12 kW. What is the COP of the unit?

Solution:

$$\frac{10 \text{ tons } 12{,}000 \text{ Btu/hr}}{\text{Ton}} = 120{,}000 \text{ Btu/hr}$$

So this is a 10-ton unit.

$$\frac{12 \text{ kW}}{10 \text{ Tons}} = 1.2 \ \frac{\text{kW}}{\text{Ton}} = \frac{3.517}{\text{COP}}$$

$$\text{COP} = \frac{3.517}{1.2} = 2.93$$

8-10

Problem: A wall has an area of 1000 ft^2 and has a thermal conductivity of 0.1 Btu/ft^2•h•°F. If there are 5500 heating degree days in the heating season, what is the total amount of heat in Btu lost through the wall during that heatging season?

Solution:

$$Q = U \times A \times HDD \times 24\ h/day$$
$$= (0.1)\ Btu\ xf\ t^2 \bullet h \bullet {}^{\circ}F\ 1000\ ft^2 \times$$
$$5500\ HDD/yr \times 24\ h/day$$

$$= (0.1) \times (1000) \times (5500) \times (24)$$
$$= 13.2 \times 10^6\ Btu/yr$$

Chapter 9

Combustion Processes and the Use of Industrial Wastes

Problem: A boiler is operating on natural gas, and is drawing 40% excess air. Using the combustion chart in Figure 9-11, find the combustion efficiency if the Stack Temperature Rise is 450 degrees F.

Solution: Using Figure 9-11, find the 40% Excess Air point near the center of the chart. Go vertically up to 450 degrees F Stack Temperature Rise and turn left horizontally to go to the Combustion Efficiency column on the left of the chart, and read 80% combustion efficiency.

Problem: Assuming an interest rate of 10% and a planning horizon of 20 years with no salvage value at the end of that time, and assuming that these analyses are being performed for a city-owned non-taxed utility, what is the economic value (i.e. present worth) of each of the alternatives on Section 9.4.2?

Given:

	Present System	Two Boilers	One Boiler
First Cost	None	$12,500,000	$14,000,000
Annual Cost			
Gas $2,500,000	$0	$0	
Coal $0	$306,900	$0	
Boiler Maintenance	$510,000	$300,000	$250,000
Waste Transportation $50,000 (40,000T × $1.25/T)	$1.25/T)	$0	$0
Waste Land filling (first year) $100,000	(40,000 T × $2.50)	$0	$0
Ash Transportation $8,700 (6940 T × $0	$8,700 (11,200 T × $0	$14,000 $1.25/T)	$1.25/T)
Ash Land filling (first year) $17,350 $0	$28,000 $0	(6940 T × $2.50/T)	(11,200 T × $2.50/T)
Annual Revenues Waste from other Companies			$450,000 (30,000 T × $15.00/T)

Hurdle rate (MARR):   10%
Project life: 20   years
Salvage value: $0
Tax rate: 0%
Landfill cost inflation (<5yrs.):   30%   /yr
Landfill cost inflation (>5yrs.):   10%   /yr
Assume no other inflation
Assume that the projects are expensed; therefore, depreciation is not a factor in the present value analysis

Solution: Compare the annual cash flows of each of the alternatives to the present system and bring the future cash flows into present day dollars.

Two Boilers

Year	Initial investment	Fuel Savings ($2,500,000 - $306,900)	Maint. Savings ($50,000 - $300,000)	Hauling Savings ($50,000 - $8,700)	Landfill Savings ($100,000 - $17,350) then adjust for landfill increases	Landfill inflation	Revenues	FV Sub-totals	[P/A, i, N] factor (1/(1 + MARR)^yr)	PV Sub-Totals
0	$ (12,500,000)							$ (12,500,000)	1	$ (12,500,000)
1		$ 2,193,100	$ (250,000)	$ 41,300	$ 82,650	30%	$0	$ 2,067,050	0.90909	$ 1,879,136
2		$ 2,193,100	$ (250,000)	$ 41,300	$ 107,445	30%	$0	$ 2,091,845	0.82645	$ 1,728,798
3		$ 2,193,100	$ (250,000)	$ 41,300	$ 139,679	30%	$0	$ 2,124,079	0.75131	$ 1,595,852
4		$ 2,193,100	$ (250,000)	$ 41,300	$ 181,582	30%	$0	$ 2,165,982	0.68301	$ 1,479,395
5		$ 2,193,100	$ (250,000)	$ 41,300	$ 236,057	30%	$0	$ 2,220,457	0.62092	$ 1,378,729
6		$ 2,193,100	$ (250,000)	$ 41,300	$ 259,662	30%	$0	$ 2,244,062	0.56447	$ 1,266,715
7		$ 2,193,100	$ (250,000)	$ 41,300	$ 285,629	10%	$0	$ 2,270,029	0.51316	$ 1,164,884
8		$ 2,193,100	$ (250,000)	$ 41,300	$ 314,191	10%	$0	$ 2,298,591	0.46651	$ 1,072,310
9		$ 2,193,100	$ (250,000)	$ 41,300	$ 345,611	10%	$0	$ 2,330,011	0.42410	$ 988,152
10		$ 2,193,100	$ (250,000)	$ 41,300	$ 380,172	10%	$0	$ 2,364,572	0.38554	$ 911,645
11		$ 2,193,100	$ (250,000)	$ 41,300	$ 418,189	10%	$0	$ 2,402,589	0.35049	$ 842,093
12		$ 2,193,100	$ (250,000)	$ 41,300	$ 460,008	10%	$0	$ 2,444,408	0.31863	$ 778,864
13		$ 2,193,100	$ (250,000)	$ 41,300	$ 506,008	10%	$0	$ 2,490,408	0.28966	$ 721,383
14		$ 2,193,100	$ (250,000)	$ 41,300	$ 556,609	10%	$0	$ 2,541,009	0.26333	$ 669,127
15		$ 2,193,100	$ (250,000)	$ 41,300	$ 612,270	10%	$0	$ 2,596,670	0.23939	$ 621,622
16		$ 2,193,100	$ (250,000)	$ 41,300	$ 673,497	10%	$0	$ 2,657,897	0.21763	$ 578,436
17		$ 2,193,100	$ (250,000)	$ 41,300	$ 740,847	10%	$0	$ 2,725,247	0.19784	$ 539,176
18		$ 2,193,100	$ (250,000)	$ 41,300	$ 814,932	10%	$0	$ 2,799,332	0.17986	$ 503,484
19		$ 2,193,100	$ (250,000)	$ 41,300	$ 896,425	10%	$0	$ 2,880,825	0.16351	$ 471,038
20		$ 2,193,100	$ (250,000)	$ 41,300	$ 986,067	10%	$0	$ 2,970,467	0.14864	$ 441,541

NPV = $ 7,132,377

One Boilers

Year	Initial investment	Fuel Savings ($2,500,000 - 0)	Maint. Savings ($50,000 - $250,000)	Hauling Savings ($50,000 - $14,000)	Landfill Savings ($100,000 - $28,000) then adjust for landfill increases	Landfill inflation	Revenues	FV Sub-totals	[P/A, i, N] factor (1/(1 + MARR)^yr)	PV Sub-Totals
0	$ (14,000,000)							$ (14,000,000)	1	$ (14,000,000)
1		$ 2,500,000	$ (200,000)	$ 36,000	$ 72,000	30%	$450,000	$ 2,358,000	0.90909	$ 2,598,182
2		$ 2,500,000	$ (200,000)	$ 36,000	$ 93,600	30%	$450,000	$ 2,879,600	0.82645	$ 2,379,835
3		$ 2,500,000	$ (200,000)	$ 36,000	$ 121,680	30%	$450,000	$ 2,907,680	0.75131	$ 2,184,583
4		$ 2,500,000	$ (200,000)	$ 36,000	$ 158,184	30%	$450,000	$ 2,944,184	0.68301	$ 2,010,917
5		$ 2,500,000	$ (200,000)	$ 36,000	$ 205,639	30%	$450,000	$ 2,991,639	0.62092	$ 1,857,573
6		$ 2,500,000	$ (200,000)	$ 36,000	$ 226,203	10%	$450,000	$ 3,012,203	0.56447	$ 1,700,310
7		$ 2,500,000	$ (200,000)	$ 36,000	$ 248,823	10%	$450,000	$ 3,034,823	0.51316	$ 1,557,344
8		$ 2,500,000	$ (200,000)	$ 36,000	$ 273,706	10%	$450,000	$ 3,059,706	0.46651	$ 1,427,375
9		$ 2,500,000	$ (200,000)	$ 36,000	$ 301,076	10%	$450,000	$ 3,087,076	0.42410	$ 1,309,222
10		$ 2,500,000	$ (200,000)	$ 36,000	$ 331,184	10%	$450,000	$ 3,117,184	0.38554	$ 1,201,809
11		$ 2,500,000	$ (200,000)	$ 36,000	$ 364,302	10%	$450,000	$ 3,150,302	0.35049	$ 1,104,162
12		$ 2,500,000	$ (200,000)	$ 36,000	$ 400,733	10%	$450,000	$ 3,186,733	0.31863	$ 1,015,391
13		$ 2,500,000	$ (200,000)	$ 36,000	$ 440,806	10%	$450,000	$ 3,226,806	0.28966	$ 934,691
14		$ 2,500,000	$ (200,000)	$ 36,000	$ 484,886	10%	$450,000	$ 3,270,886	0.26333	$ 861,327
15		$ 2,500,000	$ (200,000)	$ 36,000	$ 533,375	10%	$450,000	$ 3,319,375	0.23939	$ 794,632
16		$ 2,500,000	$ (200,000)	$ 36,000	$ 586,713	10%	$450,000	$ 3,372,713	0.21763	$ 734,001
17		$ 2,500,000	$ (200,000)	$ 36,000	$ 645,384	10%	$450,000	$ 3,431,384	0.19784	$ 678,881
18		$ 2,500,000	$ (200,000)	$ 36,000	$ 709,922	10%	$450,000	$ 3,495,922	0.17986	$ 628,772
19		$ 2,500,000	$ (200,000)	$ 36,000	$ 780,915	10%	$450,000	$ 3,556,915	0.16351	$ 583,219
20		$ 2,500,000	$ (200,000)	$ 36,000	$ 859,006	10%	$450,000	$ 3,645,006	0.14864	$ 541,807

NPV = **$ 12,104,032**

Therefore, since the one boiler system has a positive NPV and is higher than the two Boiler option, the one boiler alternative should be implemented. Please note that further sensitivity analysis should be performed.

Problem: You have the following data on the economics of a waste-burning system:

Savings (year 2000 dollars)		Costs (year 2000 dollars)	
Coal and natural gas:	$350,000/yr	Site preparation:	$335,000
Trash hauling		Building to house system:	$625,000
and landfill:	$473,000/yr	Equipment support structures:	$175,000
		Boiler and trash-handling	
		equipment:	$1,560,000
		Piping:	$275,000
		Instrumentation:	$220,000
		Crew locker room:	$175,000
		Miscellaneous mechanical	
		equipment:	$115,000
		Spare parts:	$60,000

Assume the capacity of this boiler is 28,000 lb/h. Suppose that these figures are 5 years old, that your company is contemplating the purchase of such a boiler, and that it is planned to save twice the energy amounts and have twice the capacity of the given boiler. The energy cost has been inflating at 10% per year, base construction costs have been inflating at 6% per year, the base inflation rate of the economy is 5%, and without inflation the cost of constructing a unit is $R^{0.73}$, multiplied by the cost of the existing unit, where R is the ratio between the capacity of the proposed unit and the capacity of the present unit. The combined federal and state tax rate of the company is 40%. The unit is subject to the 5 year depreciation schedule shown in Table 4-6. What is the after-tax present worth of the first 5 years of cash flows associated with this investment if the company uses a constant-dollar after-tax rate of return of 8% on this kind of investment?

Given:

	2000		2005	
boiler capacity:	28,000	lb/h	56,000	lb/h
energy savings:	350,000	/yr	700,000	/yr in 2000 dollars
cost of construction:	R^0.73			
R:	2			
cost of construction:	$3,540,000		$5,871,582	($3,540,000 x 2 ^ 0.73) in 2000 dollars
tax rate:			40%	

	2000 through 2005
energy cost inflation:	10% /yr
construction cost inflation:	6% /yr
other inflation:	5% /yr
Hurdle rate (MARR):	8%

Depreciation Schedule

Year	Depreciation
1	20.00%
2	32.00%
3	19.20%
4	11.52%
5	11.52%

Assume the construction takes a year; therefore, savings start a year after construction begins. Furthermore, that construction begins in 2005 with the recognition of costs and savings at the beginning of each year.

Solution:

Cost of Construction

Year	Inflation	Cost
2000	6%	$ 5,871,582
2001	6%	$6,223,877.33
2002	6%	$6,597,309.97
2003	6%	$6,993,148.57
2004	6%	$7,412,737.48
2005	6%	$7,857,501.73

Energy Savings

Year	Inflation	Savings
2000	10%	$ 350,000.00
2001	10%	$ 385,000.00
2002	10%	$ 423,500.00
2003	10%	$ 465,850.00
2004	10%	$ 512,435.00
2005	10%	$ 563,678.50
2006	10%	$ 620,046.35
2007	10%	$ 682,050.99
2008	10%	$ 750,256.08
2009	10%	$ 825,281.69
2010	10%	$ 907,809.86

Trash Hauling and Landfill Savings

Year	Inflation	Savings
2000	5%	$ 473,000
2001	5%	$ 496,650
2002	5%	$ 521,483
2003	5%	$ 547,557
2004	5%	$ 574,934
2005	5%	$ 603,681
2006	5%	$ 633,865
2007	5%	$ 665,558
2008	5%	$ 698,836
2009	5%	$ 733,778
2010	5%	$ 770,467

Year	Initial investment	Depreciation Cost	Energy Savings	Hauling and Landfill Savings	Before tax savings	After tax savings (before tax * (1 - 40%))	[P|A, i, N] factor (1/(1 + MARR)^yr)	PV Sub-Totals
2005	$(7,857,502)					$(7,857,502)	1	$ (7,857,502)
2006		$(1,571,500)	$ 620,046	633,865	$ (317,589)	($190,553)	0.92593	$ (176,438)
2007		$(2,514,401)	$ 682,051	665,558	$ (1,166,791)	($1,166,791)	0.85734	$ (1,000,335)
2008		$(1,508,640)	$ 750,256	698,836	$ (59,548)	($59,548)	0.79383	$ (47,271)
2009		$ (905,184)	$ 825,282	733,778	$ 653,876	$653,876	0.73503	$ 480,618
2010		$ (905,184)	$ 907,810	770,467	$ 773,093	$773,093	0.68058	$ 526,154
		$(7,404,910)					NPV =	$ (8,074,774)

Therefore, the present value of the first five years is $(8,074,774)

Problem The choice of an optimum combination of boiler sizes in the garbage-coal situation is not usually easy. Suppose health conditions limit the time garbage, even dried, can be stored to 1 month. Use initial costs given in the accompanying table, and assume the municipality and your company have supplies and needs for energy, respectively, as given in the table labeled "data" for Problem 9.4. Suppose all other costs for this problem are the same as Section 9.4.2. What is the optimum choice now?

Costs for Problem 9.4

Capacity, 750 psi	Initial Costs: Trashed-fired boiler	Initial Costs: Coal-fired boiler
50,000 lb/h	n/a	$1,800,000
100,000	n/a	$3,500,000
150,000	$6,250,000	$5,100,000
200,000	$8,640,000	$6,900,000
250,000	$10,870,000	$8,900,000
300,000	$13,000,000	$11,000,000

Month	Garbage needed (tons)	Garbage available (tons)
January	23,000	13,500
February	23,000	13,500
March	21,600	16,500
April	19,500	18,000
May	14,100	18,900
June	9,500	19,500
July	7,600	22,500
August	9,500	21,000
September	10,800	21,000
October	13,500	18,000
November	18,400	15,000
December	24,300	18,600

Assume that garbage density is 81.5 lb/cu ft and has 1015 Btu/cu ft

Given:

Hurdle rate (MARR):	10%	
Project life:	20	years
Salvage value:	$0	
Tax rate:	0%	
Landfill cost inflation (<5yrs.):	30%	/yr
Landfill cost inflation (>5yrs.):	10%	/yr

Assume no other inflation

Assume that the projects are expensed; therefore, depreciation is not a factor in the present value analysis

garbage density:	81.5	lb/cu ft
garbage energy content:	1,015	Btu/cu ft
garbage storage constraint:	1	month
hours of operation per year:	8,760	h/yr

Assume that the capacities in the above table already accounts for maintenance time and outages.

Coal energy content:	21,000,000	Btu/ton
Cost of Coal:	$55	/ton
Coal ash rate:	9.6%	
Trash ash rate:	16%	

Solution:

First we must determine the amount of garbage flow, so that the health constraint of a maximum

Month	Garbage needed (tons)	Garbage available (tons)	Old garbage (tons)	Garbage burned (tons)
December	24,300	18,600	(5,700)	18,600
January	23,000	13,500	(9,500)	13,500
February	23,000	13,500	(9,500)	13,500
March	21,600	16,500	(5,100)	16,500
April	19,500	18,000	(1,500)	18,000
May	14,100	18,900	4,800	14,100
June	9,500	19,500	14,800	9,500
July	7,600	22,500	29,700	7,600
August	9,500	21,000	41,200	9,500
September	10,800	21,000	51,400	10,800
October	13,500	18,000	55,900	13,500
November	18,400	15,000	52,500	18,400
December	24,300	18,600	46,800	24,300
January	23,000	13,500	37,300	23,000

Annual garbage available:	248,100	
Maximum garbage burned:	24,300
Minimum garbage burned:	7,600

To find the initial costs of the various options, we convert the minimum and maximum garbage to burn to an hourly capacity:

Max. Capacity = Tons/month x months/yr x lb/ton x yr/hr

= 24,300 Tons/mo x 12 mo/yr x 2,000 lb/ton x yr/8,760 hr

= 66,575.34 lb/h

Max. Capacity = Tons/month x months/yr x lb/ton x yr/hr

= 7,600 Tons/mo x 12 mo/yr x 2,000 lb/ton x yr/8,760 hr

= 20,821.92 lb/h

Therefore, this leaves us with 3 options based on the demand and equipment availability:

Equipment	Garbage Capacity (tons/mo)	Cost	Comments
150,000 lb/h trash fired	54,750	$6,250,000	Can burn all garbage in any given month
50,000 lb/h coal fired	18,250	$1,800,000	Will have left over garbage in some months
100,000 lb/h coal fired	36,500	$3,500,000	Can burn all garbage in any given month

Next, we figure out our shortages based on need by month.

Month	Garbage needed (tons)	Garbage burned (tons)	Two Boilers -- Coal fired 50,000 lb/hr tons of coal (1)	Two Boilers -- Coal fired 100,000 lb/h tons of coal (2)	One Boiler -- Trash fired Tons Garbage from other companies (3)
January	23,000	13,500	11.27	11.27	9,500
February	23,000	13,500	11.27	11.27	9,500
March	21,600	16,500	6.05	6.05	5,100
April	19,500	18,000	1.78	1.78	1,500
May	14,100	14,100	-	-	-
June	9,500	9,500	-	-	-
July	7,600	7,600	-	-	-
August	9,500	9,500	-	-	-
September	10,800	10,800	-	-	-
October	13,500	13,500	-	-	-
November	18,400	18,400	0.18	-	-
December	24,300	24,300	7.18	-	-
Annual totals:	194,800	169,200	37.72	30.36	25,600

The calculations for January for the above table are as follows:

(1)   Coal needed = (garbage needed - min(garbage burned, monthly capacity) x 2,000 lb/ton x (1/garbage density) x garbage energy content x 1/coal energy content

= (23,000 tons - min(13,500 tons,18,250 tons) x 2,000 lb/ton x (1 cu ft/81.5 lb) x 1015 Btu/cu ft x ton/21,000,000 Btu

= 11.27   tons of coal

(2)   Coal needed = (garbage needed - min(garbage burned, monthly capacity) x 2,000 lb/ton x (1/garbage density) x garbage energy content x 1/coal energy content

= (23,000 tons - min(13,500 tons,36,500 tons) x 2,000 lb/ton x (1 cu ft/81.5 lb) x 1015 Btu/cu ft x ton/21,000,000 Btu

= 11.27   tons of coal

(3)   Garbage needed other = (garbage needed - min(garbage burned, monthly capacity)

= (23,000 tons - min(13,500 tons,54,750 tons)

= 9,500.00   tons of garbage from other companies

Next, we figure out the garbage that still needs to be disposed. The only option that does not have enough capacity is the coal fired 50,000 lb/hr:

Month	Garbage burned (tons)	Capacity (tons)	Two Boilers—Coal fired 50,000 lb/hr tons of garbage left over
January	13,500	18,250	
February	13,500	18,250	
March	16,500	18,250	
April	18,000	18,250	
May	14,100	18,250	
June	9,500	18,250	
July	7,600	18,250	
August	9,500	18,250	
September	10,800	18,250	
October	13,500	18,250	
November	18,400	18,250	150
December	24,300	18,250	6,050
Annual totals:			6,200

Next, we calculate the ash waste for each option:

Month	Two Boilers— Coal fired 50,000 lb/hr	Two Boilers— Coal fired 100,000 lb/h	One Boiler— Trash fired
tons of coal	37.72	30.36	0
coal ash rate	9.6%	9.6%	9.6%
tons of trash	163,000	169,200	194,800
trash ash rate	16%	16%	16%
annual ash (tons)	26,083.62	27,074.91	31,168.00
ash hauling ($1.25/T)	$32,604.53	$33,843.64	$38,960.00
ash land fill ($2.50/T)	$65,209.05	$67,687.29	$77,920.00

annual ash = tons of coal **x** coal ash rate + tons of ash **x** trash ash rate

Next, we adjust the costs from problem 9.2 for our 3 options.

	Present System	Two Boilers -- Coal fired 50,000 lb/hr	Two Boilers -- Coal fired 100,000 lb/h	One Boiler -- Trash fired
First Cost	None	$ 1,800,000	$ 3,500,000	$ 6,250,000
Annual Cost				
Gas	$ 2,500,000	$0	$0	$0
Coal	$0	2074 37.72 tons x $55/ton	1670 30.36 tons x $55/ton	$0
Boiler Maintenance	$ 50,000	$ 300,000	$ 300,000	$ 250,000
Waste Transportation	$310,125 (248,100 T x $1.25/T)	7750 (6,200 T x $1.25 / T)	$0	$0
Waste Land filling (first year)	$620,250 (248,100T x $2.50)	15,500 (6,200 T x $2.50 / T)	$0	$0
Ash Transportation	$0	$ 32,605	$ 33,844	$ 38,960
Ash Land filling (first year)	$0	$ 65,209	$ 67,687	$ 77,920
Annual Revenues Waste from other Companies				$384,000 (25,600 T x $15.00/T)

Finally, we fill in the cash flows:

Coal 50,000 lb/hr

Year	Initial investment	Fuel Savings ($2,500,000 - $2,074)	Maint. Savings ($50,000 - $300,000)	Hauling Savings ($310,125 - $7,750 - $32,605)	Landfill Savings ($620,250 - $15,500 - $65,209) then adjust for landfill increases	Landfill inflation	Revenues	FV Sub-totals	[P/A, i, N] factor (1/(1+MARR)^yr)	PV Sub-Totals
0	$ (1,800,000)							$ (1,800,000)	1	$ (1,800,000)
1		$ 2,497,926	$ (250,000)	$ 269,770	$ 539,541	30%	$0	$ 3,057,237	0.90909	$ 2,779,306
2		$ 2,497,926	$ (250,000)	$ 269,770	$ 701,403	30%	$0	$ 3,219,099	0.82645	$ 2,660,413
3		$ 2,497,926	$ (250,000)	$ 269,770	$ 911,824	30%	$0	$ 3,429,520	0.75131	$ 2,576,649
4		$ 2,497,926	$ (250,000)	$ 269,770	$ 1,185,372	30%	$0	$ 3,703,068	0.68301	$ 2,529,245
5		$ 2,497,926	$ (250,000)	$ 269,770	$ 1,540,983	30%	$0	$ 4,058,679	0.62092	$ 2,520,120
6		$ 2,497,926	$ (250,000)	$ 269,770	$ 1,695,081	10%	$0	$ 4,212,777	0.56447	$ 2,378,003
7		$ 2,497,926	$ (250,000)	$ 269,770	$ 1,864,589	10%	$0	$ 4,382,285	0.51316	$ 2,248,805
8		$ 2,497,926	$ (250,000)	$ 269,770	$ 2,051,048	10%	$0	$ 4,568,744	0.46651	$ 2,131,353
9		$ 2,497,926	$ (250,000)	$ 269,770	$ 2,256,153	10%	$0	$ 4,773,849	0.42410	$ 2,024,578
10		$ 2,497,926	$ (250,000)	$ 269,770	$ 2,481,769	10%	$0	$ 4,999,465	0.38554	$ 1,927,510
11		$ 2,497,926	$ (250,000)	$ 269,770	$ 2,729,945	10%	$0	$ 5,247,641	0.35049	$ 1,839,266
12		$ 2,497,926	$ (250,000)	$ 269,770	$ 3,002,940	10%	$0	$ 5,520,636	0.31863	$ 1,759,045
13		$ 2,497,926	$ (250,000)	$ 269,770	$ 3,303,234	10%	$0	$ 5,820,930	0.28966	$ 1,686,116
14		$ 2,497,926	$ (250,000)	$ 269,770	$ 3,633,557	10%	$0	$ 6,151,253	0.26333	$ 1,619,817
15		$ 2,497,926	$ (250,000)	$ 269,770	$ 3,996,913	10%	$0	$ 6,514,609	0.23939	$ 1,559,546
16		$ 2,497,926	$ (250,000)	$ 269,770	$ 4,396,604	10%	$0	$ 6,914,300	0.21763	$ 1,504,753
17		$ 2,497,926	$ (250,000)	$ 269,770	$ 4,836,265	10%	$0	$ 7,353,961	0.19784	$ 1,454,942
18		$ 2,497,926	$ (250,000)	$ 269,770	$ 5,319,891	10%	$0	$ 7,837,587	0.17986	$ 1,409,659
19		$ 2,497,926	$ (250,000)	$ 269,770	$ 5,851,881	10%	$0	$ 8,369,577	0.16351	$ 1,368,493
20		$ 2,497,926	$ (250,000)	$ 269,770	$ 6,437,069	10%	$0	$ 8,954,765	0.14864	$ 1,331,069

NPV = **$ 37,508,689**

Coal 100,000 lb/hr

Year	Initial investment ($3,500,000)	Fuel Savings ($2,500,000)	Maint. Savings ($50,000 - $250,000)	Hauling Savings ($310,125 - $33,844)	Landfill Savings ($620,250 - $67,687) then adjust for landfill increases	Landfill inflation	Revenues	FV Sub-totals	[P/A, i, N] factor (1/1 + MARR)^yr)	PV Sub-totals
0	$ (3,500,000)							$ (3,500,000)	1	$ (3,500,000)
1		$ 2,500,000	$ (200,000)	$ 276,281	$ 552,563	30%	$0	$ 3,128,844	0.90909	$ 2,844,404
2		$ 2,500,000	$ (200,000)	$ 276,281	$ 718,332	30%	$0	$ 3,294,613	0.82645	$ 2,722,821
3		$ 2,500,000	$ (200,000)	$ 276,281	$ 933,831	30%	$0	$ 3,510,112	0.75131	$ 2,637,199
4		$ 2,500,000	$ (200,000)	$ 276,281	$ 1,213,981	30%	$0	$ 3,790,262	0.68301	$ 2,588,800
5		$ 2,500,000	$ (200,000)	$ 276,281	$ 1,578,175	30%	$0	$ 4,154,456	0.62092	$ 2,579,590
6		$ 2,500,000	$ (200,000)	$ 276,281	$ 1,735,993	10%	$0	$ 4,312,274	0.56447	$ 2,434,166
7		$ 2,500,000	$ (200,000)	$ 276,281	$ 1,909,592	10%	$0	$ 4,485,873	0.51316	$ 2,301,962
8		$ 2,500,000	$ (200,000)	$ 276,281	$ 2,100,551	10%	$0	$ 4,676,832	0.46651	$ 2,181,777
9		$ 2,500,000	$ (200,000)	$ 276,281	$ 2,310,606	10%	$0	$ 4,886,887	0.42410	$ 2,072,517
10		$ 2,500,000	$ (200,000)	$ 276,281	$ 2,541,667	10%	$0	$ 5,117,948	0.38554	$ 1,973,190
11		$ 2,500,000	$ (200,000)	$ 276,281	$ 2,795,834	10%	$0	$ 5,372,115	0.35049	$ 1,882,893
12		$ 2,500,000	$ (200,000)	$ 276,281	$ 3,075,417	10%	$0	$ 5,651,698	0.31863	$ 1,800,805
13		$ 2,500,000	$ (200,000)	$ 276,281	$ 3,382,959	10%	$0	$ 5,959,240	0.28966	$ 1,726,179
14		$ 2,500,000	$ (200,000)	$ 276,281	$ 3,721,255	10%	$0	$ 6,297,536	0.26333	$ 1,658,338
15		$ 2,500,000	$ (200,000)	$ 276,281	$ 4,093,380	10%	$0	$ 6,669,661	0.23939	$ 1,596,664
16		$ 2,500,000	$ (200,000)	$ 276,281	$ 4,502,718	10%	$0	$ 7,078,999	0.21763	$ 1,540,596
17		$ 2,500,000	$ (200,000)	$ 276,281	$ 4,952,990	10%	$0	$ 7,529,271	0.19784	$ 1,489,626
18		$ 2,500,000	$ (200,000)	$ 276,281	$ 5,448,289	10%	$0	$ 8,024,570	0.17986	$ 1,443,289
19		$ 2,500,000	$ (200,000)	$ 276,281	$ 5,993,118	10%	$0	$ 8,569,399	0.16351	$ 1,401,165
20		$ 2,500,000	$ (200,000)	$ 276,281	$ 6,592,429	10%	$0	$ 9,168,710	0.14864	$ 1,362,870
									NPV =	*$ 36,738,854*

Trash 150,000 lb/hr

Year	Initial investment	Fuel Savings ($2,500,000 @ $1,670)	Maint. Savings ($50,000 - $300,000)	Hauling Savings ($310,125 - $38,960)	Landfill Savings ($620,250 - $77,920) then adjust for landfill increases	Landfill inflation	Revenues	FV Sub-totals	[P/A, i, N] factor (1/(1 + MARR)^yr)	PV Sub-Totals
0	$ (6,250,000)							$ (6,250,000)	1	$ (6,250,000)
1		$ 2,498,330	$ (250,000)	$ 271,165	$ 542,330	30%	$384,000	$ 3,445,825	0.90909	$ 3,132,568
2		$ 2,498,330	$ (250,000)	$ 271,165	$ 705,029	30%	$384,000	$ 3,608,524	0.82645	$ 2,982,251
3		$ 2,498,330	$ (250,000)	$ 271,165	$ 916,538	30%	$384,000	$ 3,820,033	0.75131	$ 2,870,047
4		$ 2,498,330	$ (250,000)	$ 271,165	$ 1,191,499	30%	$384,000	$ 4,094,994	0.68301	$ 2,796,936
5		$ 2,498,330	$ (250,000)	$ 271,165	$ 1,548,949	30%	$384,000	$ 4,452,444	0.62092	$ 2,764,617
6		$ 2,498,330	$ (250,000)	$ 271,165	$ 1,703,844	10%	$384,000	$ 4,607,339	0.56447	$ 2,600,723
7		$ 2,498,330	$ (250,000)	$ 271,165	$ 1,874,228	10%	$384,000	$ 4,777,723	0.51316	$ 2,451,727
8		$ 2,498,330	$ (250,000)	$ 271,165	$ 2,061,651	10%	$384,000	$ 4,965,146	0.46651	$ 2,316,277
9		$ 2,498,330	$ (250,000)	$ 271,165	$ 2,267,816	10%	$384,000	$ 5,171,311	0.42410	$ 2,193,141
10		$ 2,498,330	$ (250,000)	$ 271,165	$ 2,494,597	10%	$384,000	$ 5,398,092	0.38554	$ 2,081,198
11		$ 2,498,330	$ (250,000)	$ 271,165	$ 2,744,057	10%	$384,000	$ 5,647,552	0.35049	$ 1,979,433
12		$ 2,498,330	$ (250,000)	$ 271,165	$ 3,018,463	10%	$384,000	$ 5,921,958	0.31863	$ 1,886,918
13		$ 2,498,330	$ (250,000)	$ 271,165	$ 3,320,309	10%	$384,000	$ 6,223,804	0.28966	$ 1,802,814
14		$ 2,498,330	$ (250,000)	$ 271,165	$ 3,652,340	10%	$384,000	$ 6,555,835	0.26333	$ 1,726,356
15		$ 2,498,330	$ (250,000)	$ 271,165	$ 4,017,574	10%	$384,000	$ 6,921,069	0.23939	$ 1,656,849
16		$ 2,498,330	$ (250,000)	$ 271,165	$ 4,419,331	10%	$384,000	$ 7,322,826	0.21763	$ 1,593,660
17		$ 2,498,330	$ (250,000)	$ 271,165	$ 4,861,265	10%	$384,000	$ 7,764,760	0.19784	$ 1,536,216
18		$ 2,498,330	$ (250,000)	$ 271,165	$ 5,347,391	10%	$384,000	$ 8,250,886	0.17986	$ 1,483,994
19		$ 2,498,330	$ (250,000)	$ 271,165	$ 5,882,130	10%	$384,000	$ 8,785,625	0.16351	$ 1,436,520
20		$ 2,498,330	$ (250,000)	$ 271,165	$ 6,470,343	10%	$384,000	$ 9,373,838	0.14864	$ 1,393,361
									NPV =	$ 36,435,608

Therefore, as far as NPV is concerned the options look very similar. Some more sensitivity analysis needs to be performed. However, the two boiler, coal/trash fired option does have the highest net present value and should be chosen assuming the sensitivity analysis holds this to be true.

Chapter 10

Steam Generation and Distribution

Problem: How much do these leaks cost per year in lost fuel?

Given: An audit of a 600-psi steam distribution system shows 50 wisps (estimated at 25 lb/h), 10 moderate leaks (estimated at 100 lb/h), and 2 leaks estimated at 750 lb/h each. The boiler efficiency is 85%, the ambient temperature is 75F, and the fuel is coal, at $65/ton and 14,500 Btu/lb. The steam system operates continuously throughout the year.

Solution: The amount of steam energy lost (q) can be estimated as follows:

q =   h x summation [number of leaks x mass flow rate of leaks]

=   (1203.7 - 43) Btu/lb x (50 leaks x 25 lb/h/leak + 10 leaks x 100 lb/h/leak + 2 leaks x 750 lb/h/leak)

=   4,352,625 Btu/h

=   38,129MMBtu/yr

Therefore, the cost (C) of these leaks can be estimated as follows:

C   =   q x $65/ton/14,500 Btu/lb/2,000 lb/ton/0.85

    =   38,129 MMBtu/yr

    =   *$ 100,543/yr*

Problem: How much heat is exchanged per pound of entering steam?

Given: Steam enters a heat exchanger at 567F and 1200 psia and
               leaves as water at 300F and 120 psia.

Solution: Look up the Enthalpy on steam tables or a Mollier diagram.

               You may need to extrapolate numbers for those not directly
               on the steam table.

               delta h =   h1 - h0
                         =   571.6 Btu / lb - 269.81 Btu / lb
                         =   *301.79 Btu/lb*

Problem: What would be the potential annual savings in the example of Section 7.5 if the amount of boiler blowdown could be decreased to an average rate of 3,000 lb/h, assuming that it remained at 400F? How much additional heat would be available from the 3,000 lb/h of blowdown water for use in heating the incoming makeup water?

Given:

Fuel cost:	$65 /ton
Energy content:	14,200 Btu/lb coal
Initial blowdown rate:	14,000 lb/h
Blowdown temperature:	400 F

Solution: hw = 375.1Btu/lb

Assuming blowdowns last for about one hour per day.

Therefore, the annual cost savings (CS) from reducing the blowdown mass flow rate can be estimated as follows:

$$
\begin{aligned}
CS1 &= \text{hw } x \text{ (m1 - m0) } x \text{ 365 h/yr } x \text{ \$65/ton/2,000 lb/} \\
&\quad \text{ton/14,200 Btu/lb} \\
&= 375.1 \text{ Btu/lb } x \text{ (14,000 lb/h - 3,000 lb/h) } x \text{ 8,760 h/yr} \\
&\quad x \text{ \$65/ton/2,000 lb/ton/14,200 Btu/lb} \\
&= \$\,3,447/yr
\end{aligned}
$$

Additionally, the heat (q) available from the 3,000 lb/h of blowdown water can be estimated as follows (assuming 100% recovery of the heat):

$$
\begin{aligned}
q &= \text{hw } x \text{ 3,000 lb/h} \\
&= 375.1 \qquad \text{Btu/lb } x \text{ 3,000 lb/h} \\
&= \textbf{1,125,300} \quad \textbf{Btu/h of blowdown}
\end{aligned}
$$

Therefore, the annual cost savings (CS2) of recovering all the heat from the blowdowns can be estimated as follows:

$$
\begin{aligned}
CS2 &= \text{q } x \text{ 365 h/yr } x \text{ \$65/ton/2,000 lb/ton/14,200 Btu/lb} \\
&= 1,125,300 \quad \text{Btu/h} \\
&\quad x \text{ 365 h/yr } x \text{ \$65/ton/2,000 lb/ton/14,200 Btu/lb} \\
&= \$940/yr
\end{aligned}
$$

Finally, the total annual cost savings (CS) from these two measures is:

$$
CS \quad = \quad \$4,387/yr
$$

Problem: Develop a table showing the size of the orifice, the number of pounds of steam lost per hour, the cost per month, and the cost for an average heating season of 7 months.

Given: Suppose that you are preparing to estimate the cost of steam leaks in a 350-psig steam system. The source of the steam is 14,200 Btu/lb coal at $70/ton, and the efficiency of the boiler plant is 70%. Hole diameters are classified as 1/16, 1/8, 1/4, 3/8, and 1/2 inches.

Solution:

Size of leak (in.)	lbm/hr lost	Heat loss (Btu/h)	Monthly cost ($)	Annual cost ($)
1/16	39	45,219	$ 116	$ 814
1/8	156	180,874	$ 465	$ 3,254
1/4	624	723,497	$ 1,860	$ 13,018
3/8	1,404	1,627,869	$ 4,184	$ 29,290
1/2	2,497	2,893,989	$ 7,439	$ 52,071

Assume starting temperature of the make-up water is room temperature (77F)

Problem: If this change is made, how many pounds per hour of steam does this Energy Management Opportunity (EMO) save?

Given: A 300-foot long steam pipe carries saturated steam at 95-psig. The pipe is not well insulated, and has a heat loss of about 50,000 Btu per hour. The plant Industrial Engineer suggests that the pipe insulation be increased so that the heat loss would be only 5,000 Btu per hour.

Solution: The heat rate savings (q) can be estimated as follows:

$$
\begin{aligned}
q &= q1 - q0 \\
&= 50{,}000 \text{ Btu/hr} - 5{,}000 \text{ Btu/hr} \\
&= 45{,}000 \text{ Btu/hr}
\end{aligned}
$$

Additionally, the mass flow rate of steam saved (m) can be calculated as follows:

$$
\begin{aligned}
m &= q/\text{hstm@95\#} \\
&= 45{,}000 \text{ Btu/hr}/1186.25 \text{ Btu/lb} \\
&= \textit{37.9 lb/h}
\end{aligned}
$$

Problem: Calculate the heat loss from the boiler from the following two sources.

Given: Tastee Orange Juice Company has a large boiler that has a 450 sq ft exposed surface that is at 225F. The boiler discharges flue gas at 400F, and has an exposed surface for the stack of 150 sq ft.

Solution:

Radiative loss   =   A x 0.1714 x 10^-8 x(Ts^4 - Tr^4)
and
Convective loss   =   A x 0.18 x (Ts - Tr)^(4/3)

where,
Ts = Surface temperature in degrees Rankine
Tr = Room temperature in degrees Rankine
A   = Surface area in square feet

Therefore, the total heat losses from these two sources can be calculated as follows:

q =   450 sq ft x [0.1714 x 10^-8 x((685R)^4 - (540R)^4)
         + 0.18 x ((685R - 540R)^(4/3))]
         + 150 sq ft x [0.1714 x 10^-8 x((860R)^4 - (540R)^4)
         + 0.18 x ((860R - 540R)^(4/3))]
   =   *343,549 Btu/h*

Problem: What is the relationship of the wisp, moderate leak, and severe leak as defined by Waterland to the hole sizes found from Grashof's formula for 600 psia steam?

In other words, find the hole sizes that correspond to the wisp, moderate leak, and severe leak.

Given: In Section 10.2.1.1 two methods were given to estimate the energy lost and cost of steam leaks.

Solution:

	Size of leak) (in. diameter	lbm/hr lost
Wisp 0.039	25	
Moderate	0.079	100
Severe	0.215	750

Assumed   600-psia saturated steam system

Chapter 11

Control Systems and Computers

Problem: How much will be saved by duty-cycling the fans such that each is off 10 minutes per hour on a rotating basis?
At any time, two fans are off and 10 are running.

How much will they be willing to spend for a control system to duty cycle the fans?

Given: Ugly Duckling Manufacturing Company has a series of 12 exhaust fans over its diagnostic laboratories. Presently, the fans run 24 hours per day, exhausting 600 cfm each. The fans are run by 2-hp motors. Assume the plant operates 24 hours per day, 365 days per year in an areas of 5,000F heating degree days and 2,000F cooling degree days per year.

The plant pays $0.05 per kWh and $5 per kW for it electricity and $5 per MMBtu for its gas. The heating plant efficiency is 0.8, and the cooling COP is 2.5. Assume the company only approves EMO projects with two year or less SPP.

Solution: Assume a load factor of 0.8 and an efficiency of 0.8 for the fans

DR fan = If × 2-hp/fan × 0.746 kW/hp × 2 fans/eff
= 0. 8 × 2-hp/fan × 0. 746 kW/hp × 2 fans/0. 8
= 2.984 kW

ES fan = DR × 24 hrs/day × 365 days/yr
= 3 kW × 24 hrs/day × 365 days/yr
= 26,140 kWh/yr

CS fan = ES × \$0.05/kWh + DR × \$5/kW/mo × 12 mo/yr
       = 26,140 kWh × \$0.05/kWh + 3 kW × \$5/kW/mo ×
       12 mo/yr
       = \$1,486/yr

heating savings = 2 fans × 600 cu ft/min/fan × 60 min/hr × 0.075
       lb/cu ft × 0.24 Btu/lb/F ×
       5,000 F days/yr × \$5/MMBtu × 24 hrs/day/0.8
       = \$972/yr

cooling savings = 2 fans × 600 cu ft/min/hr × 60 min/hr × 0.075
       lb/cu ft × 0.24 Btu/lb/F × 2,000 F days/yr ×
       1 kWh/3,412 Btu × 24 h/day × (\$0.05/kWh +
       \$5/kW/mo × 12 mo/yr/8,760 h/yr)/2.5
       = \$415/yr

Therefore, the total cost savings is:

$$CS = \$2,873/yr$$

Additionally, they would be willing to pay the following for implementation cost (IC):

$$
\begin{aligned}
IC &= SPP \times CS \\
&= 2 \text{ years} \times \$2,873/yr \\
&= \$5,745
\end{aligned}
$$

Problem: What is the savings for turning these lamps off an extra 4 hrs/day?

What type of control system would you recommend for turning off the 1,000 lamps? (manual or automatic? Timers? Other sensors?)

Given: Profits, Inc. has a present policy of leaving all of its office lights on for the cleaning crew at night. The plant closes at 6 p.m. and the cleaning crew works from 6 to 10 p.m. After a careful analysis, the company finds it can turn off 1,000 40W fluorescent lamps at closing time. The remaining 400 lamps have enough light for the cleaning crew. Assume the company works 5 days/wk, 52 wks/yr, and pays $0.06/kWh and $6/kW for electricity. Peaking hours for demand are 1 to 3 p.m. Assume there is one ballast for every two lamps and the ballast adds 15% to the load of the lamps.

Solution:

$$CS = 1{,}000 \text{ lamps} \times 40 \text{ W/lamp} \times \text{kW}/1{,}000 \text{ W} \times 4 \text{ h/day} \times$$
$$5 \text{ days/wk} \times 52 \text{ wk/yr} \times \$0.06/\text{kWh} \times 1.\,15$$
$$= \$2{,}870/yr$$

I would recommend an automatic timer for the 1,000 lamps and possibly occupancy sensors for the other 400 lamps.

Problem: How much did it cost the company in extra charges not to have the lights on some kind of control system?

What type of control system would you recommend and why?

Given: In problem 11.2, assume that the plant manager has checked on the lighting situation and discovered that the cleaning crew does not always remember to turn the remaining lights off when they leave. In the past years, the lights have been left on overnight an average of twice a month. One of the times the lights were left on over a weekend.

Solution: The cost (C) from leaving the 400 lights on 8 hours a night for 23 nights a year and 56 hours on weekend:

$$CS = 400 \text{ lamps} \times 40 \text{ W/lamp} \times \text{kW}/1{,}000 \text{ W} \times (8 \text{ h/day} \times 23 \text{ days/yr} + 56 \text{ h/yr}) \times \$0.06/\text{kWh} \times 1.\,15$$
$$= \$265/yr$$

I would recommend an automatic timer for the 1,000 lamps and possibly occupancy sensors for the other 400 lamps.

The timers would turn off the unnecessary lighting when even when the cleaning crew is working, and the occupancy sensors ensure the lights turn off when no one is present.

Problem: What is the savings in Btu for this setback?

How could this furnace setback be accomplished?

Given: Therms, Inc. has a large electric heat-treating furnace that takes considerable time to warm up. However, a careful analysis shows the furnace could be turned back from a normal temperature of 1,800F to 800F, 20 hours/week and be heated back up in time for production. The ambient temperature is 70F, and the composite R-value of the walls and roof is 12, and the total surface area is 1,000 sq ft.

Solution: q U × A × (1800F - 800F) × 20 h/wk × 52 wk/yr
        =   1/R × A × (1800F - 800F) × 20 h/wk × 52 wk/yr
        =   (1/12) Btu/(sq ft h F) × 1,000 sq ft × (1800F -800F) × 20 h/wk × 52 wk/yr
        =   *86,666,667 Btu/yr*

This furnace setback could be accomplished with a programmable logic controller (PLC).

Problem: Obtain bin data for your region, and calculate the savings in Btu for a nighttime setback of 15F from 65F to 50F, 8 hours per day (midnight to 8 a.m.)

Solution: If we assume the bin data yields 1,000 degree-days, then using Figure 11-1 the savings if the setback occurred 24 hours a day would be 225 Btu/sq ft/yr. Therefore, since 8 h/day is one-third of the time, the saving would be *75 Btu/sq ft/yr*.

Problem: What is the savings? Determine the SPP.
              Would you recommend it to the company?

Given: Petro Treatments has its security lights on timers. The com-
              pany figures an average operating time of one hour per day
              can be saved by using photocell controls. The company has
              100 mercury vapor lamps of 1,000 Watts each, and the lamp
              ballast increases the electric load by 15%. The company pays
              $0.06/kWh. Assume there is no demand savings. The pho-
              tocell controls cost $10 apiece and each lamp must have its
              own photocell. It will cost the company an average of $15
              per lamp to install the photocells.

Solution: CS = 1000 W/lamp × 100 lamps
                           × kW/1,000 W × 1.15 × 365 h/yr × $0.06/kWh
                      = *$2,519/yr*
                  SPP = IC/CS
                          ($15+$10)/lamp × 100 lamps/$2,519/yr
                          0.99 years

              Since the payback period is less than one year, I would
              recommend this project.

Problem: Would you recommend this change? Why?

Given: CKT Manufacturing Company has an office area with a
             number of windows. The offices are presently lighted with
             100 40-W fluorescent lamps. The lights are on about 3,000
             hours each year, and CKT pays $0.08 per kWh for electric-
             ity. After measuring the lighting levels throughout the office
             area for several months, you have determined that 70% of
             the lighting energy could be saved if the company installed a
             lighting system with photo sensors and dimmable electronic
             ballasts and utilized daylighting whenever possible.
             The new lighting system using 32-Watt T-8 lamps and elec-
             tronic ballasts together with the photo sensors would cost
             about $2,500.

Solution: CS = 40 W/lamp × 100 lamps × kW/1,000 W × 1. 15 ×
                  3,000 h/yr × $0.08/kWh × 0.7
                = $773/yr
             SPP = IC/CS
                = $2,500/$773/yr
                  3.23 years

             Since the payback period is less than five years, I would
             recommend this project.

Chapter 12

Maintenance

Problem: Based on this table, give a range of times for possible intervals for changing filters.

Given: In determining how often to change filters, an inclined tube manometer is installed across a filter. Conditions have been observed as follows:

Week	Manometer reading	Filter condition
1	0.4 in water	Clean
2	0.6	Clean
3	0.7	A bit dirty
4	0.8	A bit dirty
5	0.8	A bit dirty
6-9	0.9	Dirty
10-13	1.0	Dirty
14-18	1.1	Dirty
19-23	1.2	Very Dirty
24	1.3	Plugged up: changed

Solution:

One possible interval for changing the filter is once every 14 to 18 weeks.

Problem: Calculate the standard time for filter cleaning.

Given: You have been keeping careful records on the amount of time taken to clean air filters in a large HVAC system. The time taken to clean 35 filter banks was an average of 18 min/filter bank and was calculated over several days with three different people: one fast, one slow, and one average. Additional time that must be taken into account includes personal time of 20 minutes every 4 hours. Setup time was not included. Assume that fatigue and miscellaneous delay have been included in the observed times.

Solution:

$$ST = 35 \text{ filter banks} \times 18 \text{ min/filter bank} \times (1 + 20 \text{ min}/240 \text{ min})/35 \text{ filter banks}$$
$$= 19.5 \text{ min/filter}$$

Problem: How many people could you have hired for the money you lost?

Given: Your company has suffered from high employee turnover and production losses, both attributed to poor maintenance (the work area was uncomfortable, and machines also broke down). Eight people left last year, six of them probably because of employee comfort. You estimate training costs as $10,000 per person. In addition, you had one 3-week problem that probably would have been a 1-week problem if it had been caught in time. Each week cost approximately $10,000 All these might have been prevented if you had a good maintenance staff. Assume that each maintenance person costs $25,000 plus $15,000 in overhead per year.

Solution: The cost (C) due to poor maintenance conditions can be estimated as follows:

C = Six people lost due to poor maintenance (comfort) × $10,000/person in training + (3 weeks-1 week) × $10,000/week of downtime
= $80,000

Therefore, the number of maintenance people (N) you could hire can be calculated as follows:

N = C/(salary + overhead)
= $80,000/($25,000 + $15,000)/person
= *2 maintenance people*

Problem: How large an annual gas bill is needed before adding a maintenance person for the boiler alone is justified if this person would cost $40,000 per year?

Given: A recent analysis of your boiler showed that you have 15% excess combustion air. Discussion with the local gas company has revealed that you could use 5% combustion air if your controls were maintained better. This represents a calculated efficiency improvement of 2.3%.

Solution: The annual gas bill required to pay for a maintenance person that would increase boiler efficiency by 2.3% can be calculated as follows:

gas bill = $40,000/yr/2.3%
          = *$1,739,130/yr*

Problem: What annual amount would this improvement be worth considering energy costs only?

Given: Your steam distribution system is old and has many leaks. Presently, steam is being generated by a coal-fired boiler, and your coal bill for the boiler is $600,000 per year. A careful energy audit estimated that you were losing 15% of the generated steam through leaks and that this could be reduced to 2%

Solution: CS   = (15% - 2%) × $600,000/yr
              = *$78,000/yr*

Problem: Group relamping is a maintenance procedure recommended in Chapter Five. Using data from Chapter Five, construct a graph which plots maintenance cost per hour and relamping interval expressed as a percentage of the lamps rated life against the total relamping cost. Can you construct such a graph that will provide the answer to the question whether group relamping is cost-effective for a particular company?

Solution:

Maintenance Cost per hour ($) (H)	Relamping Interval of rated life (I)	Product of hourly maintenance cost and (% relamping interval) (H × I)	Relamping Cost per lamp (G)
10	75%	$7.50	$2.24
15	80%	$11.25	$2.80
20	85%	$15.00	$3.36
25		$18.75	$3.91
30		$22.50	$4.47
		$8.00	$2.10
		$12.00	$2.63
		$16.00	$3.15
		$20.00	$3.67
		$24.00	$4.19
		$8.50	$1.98
		$12.75	$2.47
		$17.00	$2.96
		$21.25	$3.45
		$25.50	$3.94

I = 80%　　　L (/lamp) = $0.85

Maintenance Cost per hour ($) (H)	Total spot relamping cost ($/lamp) (Cs)	Total group relamping cost ($/lamp) (Cg)
10	5.85	2.10
15	8.35	2.63
20	10.85	3.15
25	13.35	3.67
30	15.85	4.19

10.6 Chart 2 constructed with I=80% and lamp cost of $0.85, in which case, with the given maintenance costs, it is always cost effective to group relamp. Additional assume includes that it takes 30 minutes per lamp to spot relamp and 5 minutes per lamp to group relamp.

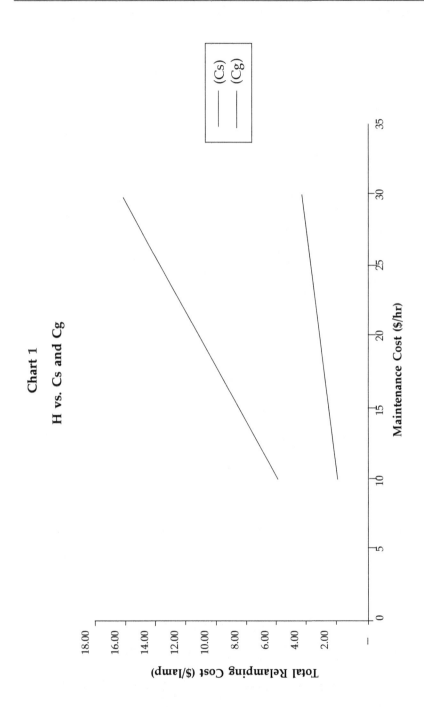

Chart 1

H vs. Cs and Cg

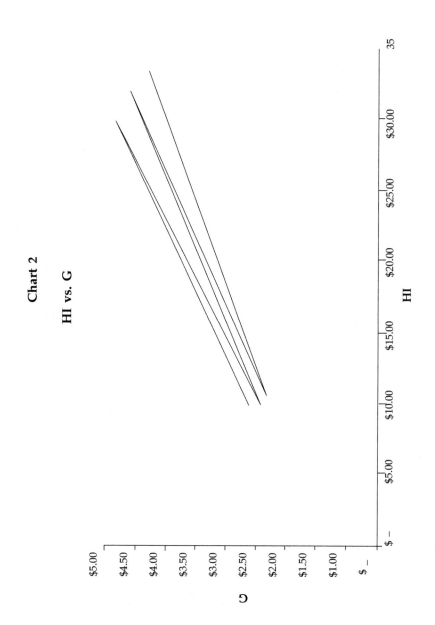

Chart 2

HI vs. G

Chapter 13

Insulation

Problem: What is the heat loss per year in Btu? What is the cost of this
            heat loss?

Given: A metal tank made out of mild steel is 4 feet in diameter, 6
            feet long, and holds water at 180F. The tank holds hot water
            all the time is on a stand so all sides are exposed to ambient
            conditions at 80F. The boiler supplying this hot water is 79%
            efficient and uses natural gas costing $5/MMBtu. Assume
            there is no air movement around the tank.

Solution: assume thickness of the tank is   0.5 inch
            K (Btu in/(sq ft h F))   314.4
            R ((sq ft h F)/Btu)   0.00159   R=d/K
            R surface   0.46   (sq ft h F)/Btu
            U   2.17   Btu/(sq ft h F)
            T(amb)   80 F
            T(inside)   180 F
            A = piDH+2pir^2 101 sq ft
            Q = UA delta T   21,779 Btu/h
            Hours per year (h)   8,760 h/yr
            q = Qh   190,786,330   Btu/yr

            c $5 /MBtu
            eff 79%

            C = qc/eff $1,208/yr

Problem: Calculate the present worth of the proposed investment.

Given: Ace Manufacturing has an uninsulated condensate return
tank holding pressurized condensate at 20 psig saturated.
The tank is 2.5 feet in diameter and 4 feet long. Management is considering adding 2 inches of aluminum-jacketed
fiberglass at an installed cost of $0.60 per sq. ft. The steam is
generated by a boiler which is 78% efficient and consumes
No. 2 fuel oil at $7/MMBtu. Energy cost will remain constant
over the economic life of the insulation of 5 years. Ambient
temperature is 70F. Rs is 0.42 for the uninsulated tank. The
tank is used 8,000 h/yr.

Solution:

assume thickness of the tank is	0.5	inch
K (Btu in/(sq ft h F))	314.4	
R ((sq ft h F)/Btu)	0.00159	R=d/K
R surface	0.42	(sq ft h F)/Btu
U	2.37	Btu/(sq ft h F)
T(amb)	70	F
T(inside)	259	F
A = piDH+2pir2	41	sq ft
Q = UA delta T	18,485	Btu/h
Hours per year (h)	8,000	h/yr
q = Qh	161,928,933	Btu/yr
c	$7	/MMBtu
eff	78%	
Co = qc/eff	$1,453	/yr

assume thickness of the tank is	0.5	inches
K (Btu in/(sq ft h F))	314.4	
R ((sq ft h F)/Btu)	0.00159	R = d/K
R surface	0.76	(sq ft h F)/Btu
K insulation	0.25	Btu in/(sq ft h F)
d insulation	2	inches
R insulation	8.00	(sq ft h F)/Btu
U	0.11	Btu/(sq ft h F)
T(amb)	70	F
T(inside)	259	F
A = piDH+2pir²	41	sq ft
Q = UA delta T	889	Btu/h
Hours per year (h)	8,000	h/yr
q = Qh	7,791,699	Btu/yr
c	$7	/MMBtu
eff	78%	
Cf qc/eff	$70	/yr

Therefore, the annual cost savings (CS) is:

CS = Co – Cf = $1,383 /yr

Additionally, the implementation cost (IC) of the insulation installation is:

IC = A × $0.60/sq ft = $24.74

Finally, the present worth (NPV) can be calculated as follows:

P   = A[P | A, 15%, 5] – IC
     = $1,156[3.3522] – $24.74
     = $4,612

Problem: What is the savings in dollars and Btu?

Given: Your plant has 500 ft of uninsulated hot water lines carrying
             water at 180F. The pipes are 4 inches in nominal diameter.
             You decide to insulate these with 2-inch calcium silicate snap-
             on insulation at $1/sq ft installed cost. The boiler supplying
             the hot water consumes natural gas at $6/MMBtu and is 80%
             efficient. Ambient air is 80F, and the lines are active 8,760 h/
             yr.

Solution:

assume thickness	0.25	inch
K (Btu in/(sq ft h F))	314.4	
R ((sq ft h F)/Btu)	0.00080	R = d/K
R surface	0.46	(sq ft h F)/Btu
U	2.17	Btu/(sq ft h F)
T(amb)	80	F
T(inside)	180	F
A = piDH	589	sq ft
Q = UA delta T	127,833	Btu/h
Hours per year (h)	8,760	h/yr
qo = Qh	1,119.82	MMBtu/yr
c	$6	/MMBtu
eff	80%	
Co = qc/eff	$8,399	/yr

assume thickness	0.25	inches
K (Btu in/(sq ft h F))	314.4	
R ((sq ft h F)/Btu)	0.00080	R = d/K
R surface	0.46	(sq ft h F)/Btu
K insulation	0.4	Btu in/(sq ft h F)
d insulation	2.70	inches
R insulation	6.76	(sq ft h F)/Btu
U	0.14	Btu/(sq ft h F)
T(amb)	80	F
T(inside)	180	F
A = piDH	1,113	sq ft
Q = UA delta T	15,415	Btu/h
Hours per year (h)	8,760	h/yr
qf = Qh	135.03	MMBtu/yr
c	$6	/MMBtu
eff	80%	
Cf = qc/eff	$1,013	/yr

Therefore, the annual cost savings (CS) is:

$$CS = Co - Cf = \$7,386/yr$$

Additionally, the amount of heat saved (ES) is:

$$ES = qo - qf = 985 \text{ MMBtu/yr}$$

Problem: What is the cost of heat loss and heat gain per sq ft for a year?

Given: Given a wall constructed as shown in Figure 11-6. EDD are 4,000 F-days, while CDD are 2,000 F-days. Heating is by gas with a unit efficiency of 0.7. Gas costs $6/MMBtu. Cooling is by electricity at $0.06/kWh (ignore demand costs), and the cooling plant has a 2.5 seasonal COP.

Solution:

Layer	R ((h sq ft F)/Btu)	d (inches)	K ((Btu in)/(h sq ft F))
Outside film	0.30		
Brick	0.44	4	9
Mortar	0.10	0.5	5
Block	0.71	4	5.6
Plaster board	0.44	0.5	1.125
inside film	0.60		
	2.60		
U (1/R)	0.38		

$$Q/A = U \times DD/yr \times 24\ h/day$$
$$Q/A\ heating = 0.38\ Btu/(h\ sq\ ft\ F) \times 4{,}000\ F\ days/yr \times 24\ h/day$$
$$36{,}878\ Btu/sq\ ft/yr$$
$$Q/A\ cooling = 0.38\ Btu/(h\ sq\ ft\ F) \times 2{,}000\ F\ days/yr \times 24\ h/day$$
$$= 18{,}439\ Btu/sq\ ft/yr$$

Therefore, the cost of the heating loss (Ch) can be estimated as follows:

Ch =   Q/A heating × $6/MMBtu/0.7
  =   36,878 Btu/sq ft/yr × $6/MMBtu/0.7
  = $0.32/sq ft/yr

Additionally, the cost of the cooling loss (Cc) can be estimated as follows:

Cc =   Q/A cooling × kWh/3,412 Btu × $0.06/kWh/2.5
  =   16,223 Btu/sq ft/yr × kWh/3,412 Btu × $0.06/kWh/2.5
  = $0.13/sq ft/yr

Finally, the total cost (C) is:

C =   Ch+Cc
  = $0.45/sq ft/yr

Problem: How much fiberglass insulation with a Kraft paper jacket is necessary to prevent condensation on the pipes?

Given: A 6-inch pipe carries chilled water at 40F in an atmosphere with a temperature of 90F and a dew point of 85F.

Solution:

Q total $= (90F-40F)/(Rp+Ri+Rs)$
          $= (90F-85F)/Rs$
          $= (85F-40F)/Ri$

Rs $= 0.5\ 3$ (h sq ft F)/Btu

Ri $= 45F \times 0.53$ (h sq ft F)/Btu/5F
      $= 4.77$ (h sq ft F)/Btu

di $= Ri \times K$
      $= 4.77$ (h sq ft F)/Btu $\times 0.25$ (Btu in)/(h sq ft F)
      $= $ *1.19 inches of fiberglass insulation*

Problem: What is the R-value of one of the walls with just a window? What is the R-value of the wall with the window and the door? What is the R-value of the roof? How many Btus must that air conditioner remove to keep the inside temperature at 78 F? How many kWh of electric energy will be used in that one hour period by the air conditioner?

Given: A building consists of fours walls that are each 8 feet high and 20 feet long. The wall is constructed of 4 inches of cork-board, with 1 inch of plaster on the outside and 1/2 inch of gypsum board on the inside. Three of the walls have 6 × 4 foot, single-pane windows with R = 0.7. The fourth wall has a 6 × 4 foot window and a 3 × 7 foot door made of one inch thick softwood. The roof is constructed of 3/4-inch plywood with asphalt roll roofing over it. The inside temperature of the building is regulated to 78F by an air-conditioner operating with a thermostat. The air-conditioner has an SEER of 8. The outside temperature is 95F for one hour.

Solution:

	Area win. (sq ft)	Area of door (sq ft)	Area of wall	Total Area less door, less win. (sq ft)
wall with 1 window	24		160	136
wall with and door	24	21	160	115
roof				400

	K ((Btu in)/(h sq ft F))	d (inches)	R = d/K ((h sq ft F)/Btu)	Wall w/o win and door	Window	Door	Roof
Corkboard	0.27	4	14.81	14.81			
Plaster	5	1	0.20	0.20			
Gypsum Board		0.5	0.44	0.44			
Surface film			0.52	0.52		0.52	0.52
Windows			0.70		0.70		
Softwood	0.8	1	1.25			1.25	
Plywood	0.8	0.75	0.94				0.94
asphalt roll roofing			0.15				0.15
				15.98	0.70	1.77	1.61

R-value of the wall with one window:

$$\text{R-value} = 1/[(1/15.98)\,(136/160) + (1/0.7)\,(24/160)$$
$$(\text{h sq ft F})/\text{Btu}]$$
$$= 3.74 \ (h \ sq \ ft \ F)/Btu$$

R-value of the wall with one window and one door:

$$\text{R-value} = 3.00 \ (h \ sq \ ft \ F)/Btu$$

R-value of the roof:

$$\text{R-value} = 1.61 \ (h \ sq \ ft \ F)/Btu$$

$$\text{Q lost} = \text{UA delta T}$$
$$= (95F - 78F) \times [(1/3.74 \ (\text{h sq ft F})/\text{Btu}) \times 160 \ \text{sq ft/wall} \times 3 \ \text{walls}$$
$$+ (1/3 \ (\text{h sq ft F})/\text{Btu}) \times 160 \ \text{sq ft/wall} \times 1 \ \text{wall}$$
$$+ (1/1.61 \ (\text{h sq ft F})/\text{Btu}) \times 400 \ \text{sq ft}]$$
$$= 7,316 \ \text{Btu/h}$$

$$q \ lost = 7,316 \ Btu \ \text{in the one hour}$$

$$kWh = 7,316 \ \text{Btu} \times \text{Wh/8 Btu} \times \text{kW/1,000 W}$$
$$= 0.914 \ kWh$$

Problem: Repeat Problem 13.6 with the single-pane windows replaced with double-paned windows have an R-value of 1.1.

Solution:

	Area win. (sq ft)	Area of door (sq ft)	Area of wall	Total Area less door, less win. (sq ft)
wall with				
1 window	24		160	136
wall with				
and door	24	21	160	115
roof				400

	K ((Btu in)/(h sq ft F))	d (inches)	R = d/K ((h sq ft F)/Btu)	Wall w/o win and door	Window	Door	Roof
Corkboard	0.27	4	14.81	14.81			
Plaster	5	1	0.20	0.20			
Gypsum							
Board		0.5	0.44	0.44			
Surface film			0.52	0.52		0.52	0.52
Windows			0.70		0.70		
Softwood	0.8	1	1.25			1.25	
Plywood	0.8	0.75	0.94				0.94
asphalt roll roofing			0.15				0.15
				15.98	0.70	1.77	1.61

R-value of the wall with one window:

R-value = $1/[(1/15.98)(136/160) + (1/1.1)(24/160)]$ (h sq ft F)/Btu

= 5.28 (h sq ft F)/Btu

R-value of the wall with one window and one door:

R-value = 3.91 (h sq ft F)/Btu

R-value of the roof:

R-value = 1.61 (h sq ft F)/Btu

Q lost = UA delta T

= (95F - 78F) × [(1/5.28 (h sq ft F)/Btu) × 160 sq ft/wall × 3 walls

+ (1/3.91 (h sq ft F)/Btu) × 160 sq ft/wall × 1 wall

+ (1/1.61 (h sq ft F)/Btu) × 400 sq ft]

= 6,468 Btu/h

q lost = 6,468 Btu in the one hour

kWh = 6,468 Btu × Wh/8 Btu × kW/1,000 W

= ***0.808 kWh***

Problem: How many dollars per year can be saved by insulating the end cap?
              What kind of insulation would you select?
              If that insulation cost $300 to install, what is the SPP for this EMO?

Given: While performing an energy audit at Ace Manufacturing Company you find that their boiler has an end cap not well insulated. The end cap is six feet in diameter and two feet long. You measure the temperature of the end cap as 250F. The temperature in the boiler room averages 90F, the boiler is used 8,760 h/yr, and fuel for the boiler is $6/MMBtu.

Solution:

assume thickness of the end cap is	2 inch	
assume the end cap is made of mild steel		
K (Btu in/(sq ft h F))	314.4	
R ((sq ft h F)/Btu)	0.00636	R=d/K
R surface	0.42	(sq ft h F)/Btu
U	2.35	Btu/(sq ft h F)
T(amb)	90 F	
T(inside)	250 F	
A = piDH+pir²	66	sq ft
Q UA delta T	24,758	Btu/h
Hours per year (h)	8,760	h/yr
q = Qh	216.88	MMBtu/yr
c	$6	/MMBtu
eff	80%	
C = qc/eff	$1,627/yr	

I would use a mineral wool fiber.

SPP = IC/C
    = $300/$1,627/yr
    = 0.18 years

Problem: What is the most cost effective solution between the two
             alternatives?

Given: Assume the tank in Problem 13. 1 is a hot water tank that
             is heated with an electrical resistance element. If this were a
             hot water tank for a residence, it would probably come with
             an insulation level of R-5. A friend says that the way to save
             money on hot water heating is to put a timer or switch on
             the tank, and to turn it off when it is not being used. Anoth-
             er friend says that the best thing to do is to put another layer
             of insulation on the tank and not turn it off and on. Assume
             that there are four of you in the residence, and that you use
             an average of 20 gallons of hot water each per day. Assume
             that you set the water temperature in the tank to 140F, and
             that the water coming into the tank is 70F. You have talked
             to an electrician, and he says that he will install a timer on
             your hot water heater for $50, or he will install an R-19 wa-
             ter heater jacket around your present water heater for $25.
             Assume that the timer can result in savings three-fourths of
             the energy lost from the water heater when it is not being
             used. Electric energy costs $0.08 per kWh.

Solution:

R ((sq ft h F)/Btu)	5	(sq ft h F)/Btu
U	0.20	Btu/(sq ft h F)
T(amb)	80	F
T(inside)	140	F
A = piDH+2pir^2	101	sq ft
Q = UA delta T	1,206	Btu/h
Hours per year (h)	8,760	h/yr
q = Qh	10,567,815	MMBtu/yr
c	$0.08	/kWh
eff	100%	
Co = qc × kWh/3,412 Btu/eff	$248	/yr

CS timer   =   75% × Co
           =   75% × $248/yr
           = $186/yr

SPP timer = $50/$185/yr
           =   0.27 years

R ((sq ft h F)/Btu)	5	(sq ft h F)/Btu
R blanket	19	(sq ft h F)/Btu
U	0.04	Btu/(sq ft h F)
T(amb)	80	F
T(inside)	140	F
A = piDH+2pir^2	101	sq ft
Q = UA delta T 251	Btu/h	
Hours per year (h)	8,760	h/yr
q = Qh	2,201,628	Btu/yr
c	$0.08	/kWh
eff	100%	
Cf = qc × kWh/3,412 Btu/eff	$52 /yr	

Therefore, the annual cost savings (CS) from the jacket is:

$$CS = Co - Cf = \$196/yr$$

SPP timer = $25/$196/yr
          =   0. 13 years

Therefore, since the jacket costs less and saves more install the jacket.

Chapter 14

Process Energy Management

14.1

Problem: If Crown Jewels buys a new motor, which one of these incentives should they ask for?

Given: Florida Electric Company offers financial incentives for large customers to replace their old electric motors with new, high efficiency motors. Crown Jewels Corporation, a large customer of FEC, has a 20-year-old 100-hp motor that they think is on its last legs, and they are considering replacing it. Their old motor is 91% efficient, and the new motor would be 95% efficient. FEC offers two different choices for incentives: either $6/hp (for the size motor considered) incentive or; a $150/kW (kW saved) incentive.

Solution: Assume a load factor of 0.6

$$\text{P saved} = 100 \text{ hp} \times 0.746 \text{ kW/hp} \times$$
$$0.6 \times [(1/0.91) - (1/0.95)]$$
$$= 2.07 \text{ kW}$$
$$\$ \text{ incentive for kW saved} = \$311$$

$$\$ \text{ incentive for size of motor} = \$600$$

Therefore, they should ask for the $6/hp (size of motor) incentive.

14.2

Problem: What is the power factor of this motor?

Given: During an energy audit at the Orange and Blue Plastics Company you saw a 100-hp electric motor that had the following information on the

nameplate: 460 v
             114 a
             3 phase
             95% efficient.

Solution:

$$P \text{ (kW)} = \text{sq rt (3)} \times v \times i \times pf \times 0.95$$

$$
\begin{aligned}
pf &= P \text{ (hp)} \times 0.746 \text{ kW/hp} / (\text{sq rt (3)} \times v \times i) / 0.95 \\
&= 100 \text{ hp} \times 0.746 \text{ kW/hp} / (\text{sq rt (3)} \times 0.460 \text{ kv} \times 114 \text{ a}) \, 0.95 \\
&= \mathbf{0.865}
\end{aligned}
$$

14.3

Problem: Using the data in Table 14-1, determine whether Ruff should purchase the high efficiency model or the standard model motor?

Find the SPP, ROI, and BCR.

Given: Ruff Metal Company has just experienced the failure of a 20-hp motor on a waste-water pump that runs about 3,000 hours a year.

Assume the new motor will last for 15 years and the company's investment rate is 15%.

Solution: Assume energy cost (EC) $0.05 /kWh
Assume demand cost (DC) $7.00 /kW/mo
Assume the motor load factor is   0.6
DR = 20 hp × 0.746 kW/hp × 0.6 × [(1/0.886) - (1/0.923)]
        = 0.41 kW

Therefore, the cost savings (CS) from using the high-efficiency motor over the standard efficiency motor can be calculated as follows:

CS = DR × DC × 12 mo/yr + DR × 3,000 h/yr × EC
     = 0.41 kW × $7/kW/mo × 12 mo/yr + 0.41 kW × 3,000 h/yr × $0.05/kWh
     = $94.78/yr

SPP = Cost premium/CS
       = $186/$94.78/yr
       = *1.96 years*

ROI = *50.85%*

BCR = PV benefits/PV cost
       = $554.21/$186
       = *2.98*

Therefore, buying the high efficiency motor seems to be a good investment.

14.4

Problem: How would you estimate the amount of waste heat that could be recovered for use in heating wash water for metal parts?

Given: A rule of thumb for an air compressor is that only 10% of the energy the air compressor uses is transferred into the compressed air. The remaining 90% becomes waste heat. You have seen a 50-hp air compressor on an audit of a facility, but you do not have any measurements of air flow rates or temperatures.

Solution: Assume that the motor efficiency is 91.5%.
Assume that the compressor motor load factor is 0.6.
Additionally, assume that 80% of the waste heat can be recovered. Therefore, one can calculate the amount of waste heat available as follows:

$$
\begin{aligned}
Q &= \text{lf} \times \text{P} \times 0.746 \text{ kW/hp} \times 90\% \times 80\% \times \\
  &\quad 3{,}412 \text{ Btu/kWh/91.5\%} \\
  &= 0.6 \times 50 \text{ hp} \times 0.746 \text{ kW/hp} \times 90\% \times 80\% \times \\
  &\quad 3{,}412 \text{ Btu/kWh/91.5\%} \\
  &= \boldsymbol{60{,}087 \text{ } Btu/h}
\end{aligned}
$$

14.5

Problem: How much would this load shifting save Orange and Blue
               Plastics on their annual electric costs?

Given: Orange and Blue Plastics has a 150-hp fire pump that must
               be tested each month to insure its availability for emergency
               use. The motor is 93% efficient, and must be run 30 min-
               utes to check its operations. The facility pays $7/kW for
               its demand charge and $0.05/kWh for energy. During your
               energy audit visit to Orange and Blue, you were told that
               they check out the fire pump during the day (which is their
               peak time), once a month. You suggest that they pay one of
               the maintenance persons an extra $50 a month to come in
               one evening a month to start up the fire pump and run it
               for 30 minutes.

Solution:

               Assume the motor load factor is 0.6

$$DR = 150 \text{ hp} \times 0.746 \text{ kW/hp} \times 0.6 \times 1/0.93$$
$$= 72.19 \text{ kW}$$

               Therefore, the electric cost savings (CS) from load shifting
               can be calculated as follows:

$$CS = DR \times DC \times 12 \text{ mo/yr}$$
$$= 72.19 \text{ kW} \times \$7/\text{kW/mo} \times 12 \text{ mo/yr}$$
$$= \$6,064/yr$$

14.6

Problem: What is the implied efficiency of a motor if we say its load
is 1 kW per hp?

What is the implied COP of an air conditioner that has a
load of 1 kW per ton?

Given: Our "rules of thumb" for the load of a motor and air conditioner have implicit assumptions on their efficiencies.

Solution:

eff = 0.746 kW/hp × 1 hp/kW
     = 74.6%
COP = 1 ton/kW × 12,000 Btu/tonh × kWh/3,412 Btu
     = 3.52

14.7

Problem: Even though the motor is expected to last another five years, you think that the company might be better off replacing the motor with a new high-efficiency model. Provide an analysis to show whether this is a cost-effective suggestion.

Given: During an audit trip to a wood products company, you note that they have a 50-hp motor driving the dust collection system. You are told that the motor is not a high efficiency model, and that it is only 10 years old. The dust collection system operates 6,000 hours each year.

Solution:

Since cost premiums range from 10% to 30%, we use 20% to calculate the cost of the high efficiency motor from the premium column in Table 12-1. Therefore, the cost of a 50-hp high-efficiency motor is 5 times $469:

$2,345

$$\begin{aligned}
\text{Assume energy cost (EC)} \quad &\$0.05 \quad /\text{kWh} \\
\text{Assume demand cost (DC)} \quad &\$7.\,00 \quad /\text{kW}/\text{mo} \\
\text{Assume the motor load factor is} \quad &0.6 \\
\text{DR} \;=\; &50 \text{ hp} \times 0.746 \text{ kW/hp} \times 0.6 \times [(1/0.915) - (1/0.938)] \\
\;=\; &0.60 \text{ kW}
\end{aligned}$$

Therefore, the cost savings (CS) from using the high-efficiency motor over the standard efficiency motor can be calculated as follows:

$$\begin{aligned}
\text{CS} \;=\; &\text{DR} \times \text{DC} \times 12 \text{ mo/yr} + \text{DR} \times 6{,}000 \text{ h/yr} \times \text{EC} \\
\;=\; &0.6 \text{ kW} \times \$7/\text{kW}/\text{mo} \times 12 \text{ mo/yr} + 0.6 \text{ kW} \times \\
&6{,}000 \text{ h/yr} \times \$0.05/\text{kWh} \\
\;=\; &\$230.30/\text{yr}
\end{aligned}$$

$$\begin{aligned}
\textit{SPP} \;=\; &\text{Cost/CS} \\
\;=\; &\$2{,}345/\$230.30/\text{yr} \\
\;=\; &\textit{10.\,2 years}
\end{aligned}$$

$$\begin{aligned}
\textit{ROI} \;=\; &\textit{5.28\%} \\
\textit{NPV} \;=\; &\textit{(\$998)} \text{ assuming a MARR of 15\%}
\end{aligned}$$

Therefore, it seems to be a bad project to change the motor now.

Chapter 15

Renewable Energy Sources and Water Management

Problem: How many gallons of water would be required to store 1 MMBtu?

Given: In designing a solar thermal system for space heating, it is determined that water will be used as a storage medium. Assuming the water temperature can vary from 80F up to 140F.

Solution:

$$Q = MC \text{ (delta T)}$$
$$M = Q/C \text{ (delta T)}$$
$$= 1 \text{ MMBtu}/(1 \text{ Btu/lb/F} \times (140 \text{ F} - 80 \text{ F}))$$
$$= 16,667 \text{ lb} \quad (8.34 \text{ lb/gal})$$
$$= \textit{1,998 gal}$$

Problem: Design the necessary array but neglect any voltage-regulating or storage device.

Given: In designing a system for photovoltaics, cells producing 0.5 volts and 1 ampere are to be used. The need is for a small dc water pump. Drawing 12 volts and 3 amperes.

Solution:

Three branches with 24 cells in each branch:

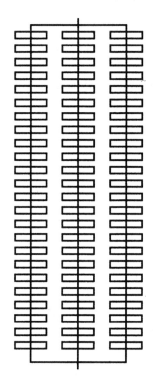

Problem: Calculate the annual water savings (gallons and dollars) and annual energy savings (MMBtu and dollars) if the water could be used as boiler makeup water.

Given: A once-through water cooling system exists for a 100-hp air compressor. The flow rate is 3 gpm. Water enters the compressor at 65F and leaves at 105F. Water and sewage cost $1.50/1,000 gallons and energy costs $5/MMBtu. Assume the water cools to 90F before it can be used and flows 8,760 h/yr.

Solution: Assume the efficiency of the heating system is 70%.

$$V = 3 \text{ gal/min} \times 60 \text{ min/h} \times 8,760 \text{ h/yr}$$
$$= 1,576,800 \text{ gal/yr}$$
$$m = 13,150,512 \text{ lb/yr } (8.34 \text{ lb/gal})$$

$$CS = 1,576,800 \text{ gal/yr} \times \$1.50/1,000 \text{ gal} + 13,150,512 \text{ lb/yr} \times$$
$$1 \text{ Btu/lb/F} \times (90F\text{-}65F) \times \$5/\text{MMBtu } 0.7$$
$$= \$4,714/yr$$

Problem: What is the net annual savings if the sawdust is burned?

Given: A large furniture plant develops 10 tons of sawdust (6,000 Btu/ton) per day that is presently hauled to the landfill for disposal at a cost of $10/ton. The sawdust could be burned in a boiler to develop steam for plant use. The steam is presently supplied by a natural gas boiler operating at 78% efficiency. Natural gas costs $5/MMBtu. Sawdust handling and in-process storage costs for the proposed system would be $3/ton. Maintenance of the equipment will cost an estimated $10,000 per year. The plant operates 250 days/yr

Solution: Assume the efficiency of the sawdust burning boiler is 70% of the efficiency of the natural gas burning boiler.

$$m = 10 \text{ tons/day} \times 250 \text{ day/yr}$$
$$= 2,500 \text{ tons/yr}$$

$$CS = 2,500 \text{ tons/yr} \times 0.7 \ (6,000 \text{ Btu/ton} \times \$5/\text{MMBtu} + (\$10/\text{ton} - \$3/\text{ton})) - \$10,000/\text{yr}$$
$$= \$2,303/yr$$

Problem: At 40 degree N latitude, how many square feet of solar
            collectors would be required to produce each month of the
            energy content of
            a) one barrel of crude oil?
            b) one ton of coal?
            c) one therm of natural gas?

Solution: Using Table 13-1, look up the data for 40 degree N latitude
            averages:

            0.6 MMBtu/sq ft/yr

            Assume that the efficiency of the cell is 0.7 the efficiency of
            the present fuel

            a) A = 5, 100,000 Btu/barrel of crude oil/mo ×
                    12 mo/yr/0. 6 MMBtu/sq ft/yr 0. 7
                  = *146 sq ft*

            b) A = 25, 000, 000 Btu/ton of coal/mo ×
                    12 mo/yr/0.6 MMBtu/sq ft/yr/0. 7
                  = *714 sq ft*

            c) A = 100,000 Btu/therm of nat. gas/mo ×
                    12 mo/yr/0.6 MMBtu/sq ft/yr/0.7
                  = *2.9 sq ft*

Problem: Determine whether Portland, OR; New Orleans, LA; or Boston, MA, have the greatest amount of solar energy per square foot of collector surface?

Given: Use Table 13.1. Assume each collector is mounted at the optimum tilt angle for that location.

Solution:

city Portland	Slope	Average Daily Radiation (Btu/day/sq ft)												Total (Btu/yr/sq ft)
		Jan	Feb	Mar	Apr	May	Jun	Jul	Aug	Sep	Oct	Nov	Dec	
	hor	578	872	1321	1495	1889	1992	2065	1774	1410	1005	5780	508	
	30	1015	1308	1684	1602	1836	1853	1959	1830	1670	1427	941	941	
	40	1114	1393	1727	1569	1746	1739	1848	1771	1680	1502	1020	1042	
	50	1184	1442	1727	1502	1622	1594	1702	1673	1651	1539	1073	1116	
	vert	1149	1279	1326	953	889	824	890	989	1172	1309	1010	1109	
Average Monthly Radiation		9E+05	8E+05	1E+06	1E+06	1E+06	1E+06	2E+06	1E+06	1E+06	1E+06	4E+06	8E+05	17,372,304

City New Orleans	Slope	Average Daily Radiation (Btu/day/sq ft)												Total (Btu/yr/sq ft)
		Jan	Feb	Mar	Apr	May	Jun	Jul	Aug	Sep	Oct	Nov	Dec	
	hor	788	954	1235	1518	1655	1633	1537	1533	1411	1316	1024	729	
	30	1061	1162	1356	1495	1499	1428	1369	1456	1490	1604	1402	1009	
	40	1106	1182	1339	1424	1389	1309	1263	1371	1451	1626	1464	1058	
	50	1125	1174	1292	1324	1256	1170	1137	1259	1381	1610	1490	1082	
	vert	944	899	847	719	599	546	548	647	843	1189	1240	929	
Average Monthly Radiation		8E+05	8E+05	1E+06	1E+06	1E+06	1E+06	1E+06	1E+06	1E+06	1E+06	1E+06	8E+05	12,584,640

City Boston	Slope	Average Daily Radiation (Btu/day/sq ft)												Total (Btu/yr/sq ft)
		Jan	Feb	Mar	Apr	May	Jun	Jul	Aug	Sep	Oct	Nov	Dec	
	hor	511	729	1078	1340	1738	1837	1826	1565	1255	876	533	438	
	30	830	1021	1313	1414	1677	1701	1722	1593	1449	1184	818	736	
	40	900	1074	1333	1379	1592	1595	1623	1536	1450	1234	878	803	
	50	947	1101	1322	1316	1477	1461	1494	1448	1417	1254	916	850	
	vert	895	950	996	831	810	759	790	857	993	1044	842	820	
Average Monthly Radiation		7E+05	7E+05	1E+06	1E+06	1E+06	1E+06	1E+06	1E+06	1E+06	9E+05	7E+05	6E+05	11,882,616

Therefore, Portland, OR, has greatest amount of solar energy per square foot of collector surface.

Problem: How many gallons of gasoline is this?
Using the maximum Btu contents shown in Table 13-15, how many pounds of corn cobs would it take to equal the Btus needed to run the car for one year? Rice hulls? Dirty solvent?

Given: A family car typically consumes about 70 million Btu per year in fuel.

Solution:

$$V = 70,000,000 \text{ Btu/yr} \times 1 \text{ gal gasoline/125,000 Btu}$$
$$= \mathbf{560 \ gal/yr}$$

Corn cobs:
$$m \quad 70,000,000 \text{ Btu/yr} \times \text{lb/5,850 Btu}$$
$$= \mathbf{11,966 \ lb/yr}$$

Rice hulls:
$$m \quad 70,000,000 \text{ Btu/yr} \times \text{lb/8,150 Btu}$$
$$= \mathbf{8,589 \ lb/yr}$$

Dirty solvent:
$$m = 70,000,000 \text{ Btu/yr} \times \text{lb/13,000 Btu}$$
$$= \mathbf{5,385 \ lb/yr}$$

Problem: Determine the power outputs in Watts per square foot for a
              good wind site and an outstanding wind site as defined in
              Section 15.5.

Solution:

$$P/A = 0.5 \times \text{density of air} \times \text{velocity}^2$$
$$= K(\text{velocity}^3)$$
$$K = 5.08 \times 10^3$$

Good site: V   13 mi/h
         P/A $= 5.08/1000 \times (13)^3$
              $= \textbf{11.16 W/sq ft}$

Outstanding site: V   19 mi/h
              P/A $= 5.08/1000 \times (13)^3$
                   34.84 W/sq ft

Problem: How much difference—in percent—is there between the
two sites in Problem 15-9?

Solution:

$$P/A = 0.5 \times \text{density of air} \times \text{velocity}^2$$
$$= K(\text{velocity}^3)$$
$$K = 5.08 \times 10^3$$

Good site: V $= 13$ mi/h
$$P/A = 5.08/1000 \times (13)^3$$
$$= 11.16 \text{ W/sq ft}$$

Outstanding site: V $= 19$ mi/h
$$P/A = 5.08/1000 \times (13)^3$$
$$= 34.84 \text{ W/sq ft}$$

% difference $=$ (outstanding - good)/outstanding
$$= 68\%$$

or

% difference $=$ (outstanding - good)/good
$$= 212\%$$

Supplemental

Problem: What is the load factor?

Given: A three-phase 50-hp motor draws 27 amps at 480 volts.

   It is 92% efficient and has a reactive power of 10 kVAR.

Solution:

Apparent power $= (3^{0.5}) \times v \times i$
   $= (3^{0.5}) \times 480 \text{ V} \times 27 \text{ amps}$
   $= 22.45 \text{ kVA}$

Reactive power $= 10 \text{ kVAR}$

sin (theta) $= 10 \text{ kVAR}/22.45 \text{ kVA}$
theta $= 0.462$

pf $= \cos \text{ (theta)}$
   $= 0.895$

Real power $= (3^{0.5}) \times v \times i \times pf$
   $= 20.10 \text{ kW}$ which is the power actually used

Rated power $= 50 \text{ hp} \times 0.746 \text{ kW/hp}/0.92$
   $= 40.54 \text{ kW}$

Load factor $= 20.1 \text{ kW}/40,54 \text{ kW}$
   $= \mathbf{49.6\%}$

Printed in the United States
by Baker & Taylor Publisher Services